Fundamentals of Cellular Network Planning and Optimisation

Fundamentals of Cellular Network Planning and Optimisation

2G/2.5G/3G... Evolution to 4G

Author:

Ajay R. Mishra

John Wiley & Sons, Ltd

Other Wiley Editorial Offices

John Wiley & Sons Inc., 111 River Street, Hoboken, NJ 07030, USA

Jossey-Bass, 989 Market Street, San Francisco, CA 94103-1741, USA

Wiley-VCH Verlag GmbH, Boschstr. 12, D-69469 Weinheim, Germany

John Wiley & Sons Australia Ltd, 33 Park Road, Milton, Queensland 4064, Australia

John Wiley & Sons (Asia) Pte Ltd, 2 Clementi Loop #02-01, Jin Xing Distripark, Singapore 129809

John Wiley & Sons Canada Ltd, 22 Worcester Road, Etobicoke, Ontario, Canada M9W 1L1

Wiley also publishes its books in a variety of electronic formats. Some content that appears in print may not be
available in electronic books.

Library of Congress Cataloging-in-Publication Data

British Library Cataloguing in Publication Data

A catalogue record for this book is available from the British Library

ISBN-10 0-470-86267-X (H/B)
ISBN-13 978-0-470-86267-4 (H/B)

Typeset in 10/12pt Times by TechBooks Electronic Services, New Delhi, India
Printed and bound in Great Britain by Antony Rowe, Chippenham Wiltshire
This book is printed on acid-free paper responsibly manufactured from sustainable forestry
in which at least two trees are planted for each one used for paper production.

Dedicated
to
The Lotus Feet of my Guru

Contents

Fundamentals of Cellular Network Planning & Optimisation A.R. Mishra.
© 2004 John Wiley & Sons, Ltd. ISBN: 0-470-86267-X

Foreword

The ever evolving desire for us all to have, at our fingertips, the immediate ability to see, hear, interact, speak, read and write with all manner of other parties, wherever we are, is one of the most remarkably rapid evolutions ever. It seems that the more we are able to enrich our lives through this enhanced mobility, then the more we demand it. Yesterday's aspirations for mobility have clearly become today's fundamental needs and this trend is strongly set to continue.

Invisible to the vast majority of habitual consumers who enjoy these mobile applications and services, are some of the most complex machines that man has ever made. The networks that have enabled the rampant expansion of mobile telephony during the 1990s continue to expand and are now evolving through the rapid data transmission, to the full video interaction that was the stuff science fiction within the time span of just one generation.

These networks are the result of some of the most intense research and development programmes, which have generated tremendously advanced products that constitute the essential building blocks of these networks. The knitting together of these elements into a fully functioning optimised and high quality network that allows the best possible user experience is a technically demanding and complex discipline.

As the network planning teams that design and optimise these networks are forever at the leading edge of the technology they are dealing with, there is a strong imperative for them to both fully understand the engineering process itself as well as the technology as it develops. This need stretches across all aspects of the network from the radio access, through the transmission and to the core. It also straddles all the technologies from GSM, through GPRS to the WCDMA networks and beyond to 4G.

Ajay R. Mishra is a network planning practitioner. He has spent his career since graduation in this discipline and understands fully the needs of the planning teams. He has used his practical insight to bring together, under one cover, the end to end view of network

Fundamentals of Cellular Network Planning & Optimisation A.R. Mishra.
© 2004 John Wiley & Sons, Ltd. ISBN: 0-470-86267-X

planning and addressed this across the technologies of GSM, GPRS, EDGE and WCDMA. The book is designed for network planners who are starting their careers in the field and for others who desire an end-to-end understanding. It emphasises the planning process itself and develops the concepts that planners need to get good results. It is not targeted at expert radio planners who want to know more about radio planning, rather it brings to them the benefit of understanding more about core or transmission planning – and likewise for all the planning disciplines. Through this book Ajay has captured what is needed to both give new planners a flying start and experienced planners a broader view.

Anthony Crane
Vice-President, Nokia Networks

Helsinki,
Finland

As operators around the world are offering an ever-increasing variety of mobile services, the underlying networks require continuous, meticulous attention, to provide highest performance and asset utilization at lowest operating cost. Dealing with this technology, that literally is transforming our patterns of communication, if not the way we conduct our lives, is strongly driving our best and brightest young engineers.

In this book Ajay R. Mishra, a highly accomplished network-planning engineer himself, has for the first time laid out the complete introduction for the engineer entering the practical domain of mobile network design and optimisation. In a compact format Ajay goes beyond the beaten path of describing technology, and establishes the practise of conceiving and optimising radio, transmission and core subsystems of today's mobile networks. The concentration on GSM and its descendants allows him to create a coherent, dependable guide for both the beginner and the field proven, senior engineer who wants to gain insight into the neighbouring domains of network design.

With operational efficiency being one of the key drivers of commercial success for mobile operators today, this compact guide's focus on proven process helps practitioners to hone their own skills. Its complete coverage of radio, transmission and core aspects allows the reader to gain full appreciation of end-to-end network performance. As mobile data services rapidly gain popularity, solid understanding of their demands on the underlying networks becomes a must for any mobile engineer. It is this broad perspective, combined with a very practical approach, which make this book stand out on its own.

Klaus Goerke
General Manager (OS)
Nokia Networks

Helsinki,
Finland

Preface

The last few years have witnessed a phenomenal growth in the wireless industry, both in terms of mobile technology and its subscribers. Both the mobile network operators and vendors have felt the importance of efficient networks with equally efficient design. This has resulted in Network Planning and Optimisation related services coming into focus. However, with the technological advances and the simultaneous existence of the 2G/2.5G/3G networks, the impact of services on network efficiency have become even more critical. Many more designing scenarios have developed with not only 2G networks but also with the evolution of 2G to 2.5G or even to 3G networks. Along with this, inter-operability of the networks has to be considered.

With scores of books available on radio communications and technology, the real need for the planning engineers remained un-fulfilled. This book fills the gap between mobile technology and mobile networks. Books that are available on system engineering deal with radio planning aspects only. This book covers radio, transmission and core at the same time. Apart from the planning and optimisation aspects for GSM, GPRS, EDGE, WCDMA for all three sections (radio/transmission/core) are also present under the same cover. This will help planning & optimisation engineers to understand not only the technological evolution but also the changes and modifications taking place in the planning and optimisation processes.

Due to these reasons, issues related to the network planning and optimisation for each of the generation of networks is dealt with. This book has been divided into four parts. Each of these parts deals with 2G, 2.5G, 3G and 4G networks. There are ten chapters in this book. Part I, Part II and Part III deal with 2G (GSM), 2.5G (GPRS and EDGE) and 3G (WCDMA) networks respectively. All these three parts deal with fundamentals of Radio, Transmission, Core network planning and optimisation. However, Part I and Part III have three chapters, each dealing exclusively with planning and optimisation of these three sections of the

Fundamentals of Cellular Network Planning & Optimisation A.R. Mishra.
© 2004 John Wiley & Sons, Ltd. ISBN: 0-470-86267-X

network. Part II contains two chapters focussing on GPRS and EDGE network planning and optimisation.

Part I contains three chapters that focus on 2G-GSM Network planning and optimisation.

Chapter 1 deals with the an overview of mobile networks. It deals with a brief history of mobile networks and its evolution. Concepts from information theory are explained briefly which are related to mobile network that would be required by network planning engineers. In conclusion an overview of the two main networks i.e. GSM and WCDMA is given.

Chapter 2 focuses on GSM radio network planning and optimisation. That scope of radio network planning is explained along with some basic concepts that are required by radio network planning engineers to plan an efficient radio network. After this, process of radio network planning is explained followed by pre-planning and detail planning aspects. In conclusion, radio network optimisation process is explained.

Chapter 3 and Chapter 4 follow a similar approach for Transmission and Core Network Planning. Both of these chapters cover scope, pre-planning, detail planning and optimisation for transmission and core networks respectively.

Part II focuses on 2.5G- GPRS and EDGE network planning and optimisation

Chapter 5 focuses on GPRS and Chapter 6 focuses on EDGE Networks. Both these chapters introduce the GPRS and EDGE Networks respectively followed by the planning processes and optimisation for radio, transmission and core networks.

Part III contains three chapters and focuses on the 3G-WCDMA Network Planning and Optimisation.

Chapter 7, 8 and 9 deal with the radio, transmission and core networks respectively. The structure of these chapters is similar to that of chapter 2, 3 and 4 respectively.

Part IV gives an overview of fourth generation networks.

Though fourth generation networks are still a distant reality from a mobile subscriber perspective, it still needs some attention from people guiding the way of the mobile world. Hence, few pages are devoted in order to give a very brief overview of what is in the store for planning engineers.

A number of appendices are also given. These appendices contributed by the experts in the respective fields deal with aspects such as planning tools and hot topics such as multimedia planning and optimisation, location based services, E2E performance measurements etc . . . The Erlang B tables given will come in handy to planning engineers in their day-to-day work.

In conclusion, there are a list of carefully chosen books and papers given that I'm sure , readers will find useful.

Finally, all suggestions and comments for the improvement of this book are welcome. Readers are requested to mail their comments to: FCNP@hotmail.com.

Acknowledgements

I fall short of words to thank everyone who has helped in bringing this book to completion.

My first thanks go to Matti Makkonen who encouraged me to go ahead with the idea of writing a book on network planning. My gratitude also goes to Veli-Pekka Somila for boosting my efforts and enquiring from time to time about the progress of the book.

Big thanks go to Tomas Novosad, for guiding me and encouraging me to write.

Thanks are due to the following colleagues and friends, for taking the time to read the manuscript and give valuable comments: Silvia Bertozzi, Sheyam Dhomeja, Ville Ruikka, Jussi Viero, Sonia Barriero, Maria Sanchez, Bogdan Cosmin, Farida Abbes, Guilherme Pizzato, Mika Sarkioja and S.K. Acharya.

Thanks also go to Azizah Aziz for helping me during the last phases of writing.

Many thanks to Anthony Crane and Klaus Goerke for donating their time and sharing with us their vision about the mobile industry in the foreword section of this book.

I would also like to thank Olli Nousia, Asok Sit and Juha Sarkioja for their support and guidance in the past years.

Ari Niininen, Christophe Landemaine, Carlos Crespo, Johanna Kahkonen, N. B. Kamat and Nezha Larhissi provided the materials in the appendices, at short notice, for which I am grateful.

Thanks go to the team at John Wiley & Sons, with special mentions for Sophie Evans, Mark Hammond and Daniel Gill who gave me excellent support throughout this project, and especially for their patience.

Finally I should like to thank my parents, Mrs Sarojini Devi Mishra and Mr Bhumitra Mishra, who gave me the inspiration to undertake this project and deliver it to the best of my capability.

Fundamentals of Cellular Network Planning & Optimisation A.R. Mishra.
© 2004 John Wiley & Sons, Ltd. ISBN: 0-470-86267-X

Introduction

The past few years have witnessed a phenomenal growth in the wireless industry, both in terms of mobile technology and subscribers. Mobile network operators and vendors have recognized the importance of efficient networks with equally efficient design processes. This has resulted in services related to network planning and optimization coming into sharp focus.

With all the technological advances, and the simultaneous existence of 2G, 2.5G and 3G networks, the impact of network efficiency has become even more critical. Many new designing scenarios have developed, and the inter-operability of the networks has to be considered.

Although scores of books are available on radio communications and technologies, the real need for a book for planning engineers remained unfulfilled. This book aims to cover the gap between mobile technologies and mobile networks. While earlier books on system engineering deal with radio planning aspects only, this book covers radio, transmission and core networks at the same time. Planning and optimization aspects for GSM, GPRS, EDGE and WCDMA, for all three sections (radio/transmission/core), are presented together. This will help planning and optimization engineers to understand not only the technological evolution but also the changes and modifications taking place in planning and optimization processes.

The book is divided into four parts. After an introductory chapter, Part I deals with 2G (GSM) networks. Then Parts II and III deal with 2.5G (GPRS and EDGE) and 3G (WCDMA) networks respectively. All these three parts deal with fundamentals of radio, transmission and core network planning and optimization. Part IV of the book looks to the future: fourth-generation networks.

Chapter 1 is an overview of mobile networks. It gives a brief history of mobile networks and their evolution. Some concepts from information theory are explained briefly. The chapter also has an overview of the two main networks, GSM and WCDMA.

Fundamentals of Cellular Network Planning & Optimisation A.R. Mishra.
© 2004 John Wiley & Sons, Ltd. ISBN: 0-470-86267-X

Chapter 2 focuses on GSM radio network planning and optimization. The scope of radio network planning is explained along with some basic concepts that are required by engineers in the planning of an efficient radio network. After this, the process of radio network planning is explained, followed by pre-planning and detailed planning aspects. The optimization process is then explained. Chapters 3 and 4 follow a similar approach for transmission and core network planning for GSM.

Moving on to 2.5G networks, Chapter 5 focuses on GPRS and Chapter 6 on EDGE. Both these chapters describe the networks, as well as planning processes and optimization for their radio, transmission and core networks.

Moving on to 3G networks, Chapters 7, 8 and 9 deal with their radio, transmission and core networks respectively. The structure of these chapters is similar to that of the earlier chapters.

Finally, Chapter 10 gives an overview of fourth-generation networks. Although 4G networks are still a distant reality from the perspective of mobile subscribers, still the subject needs some attention from people guiding the way of the mobile world. The chapter shows what is in store for the planning engineers.

There are five appendices. These have been contributed by experts in their respective fields, dealing with subjects such as planning tools, and hot topics such as multi–media planning and optimization, location-based services, end-to-end (E2E) performance measurements etc. Erlang B tables are presented as an aid to planning engineers in their day-to-day work.

Finally there is a list of recommended reading in the Bibliography.

1

Overview of Mobile Networks

1.1 INTRODUCTION

Mobile networks are differentiated from each other by the word 'generation', such as first-generation, 'second-generation' etc. This is quite appropriate because there is a big 'generation gap' between the technologies.

The first-generation mobile systems were the analogue (or semi-analogue) systems, which came in the early 1980s – they were also called NMT (Nordic Mobile Telephone). They offered mainly speech and related services and were highly incompatible with each other. Thus, their main limitations were the limited services offered and incompatibility.

The increasing necessity for a system catering for mobile communication needs, and offering more compatibility, resulted in the birth of the second-generation mobile systems. International bodies played a key role in evolving a system that would provide better services and be more transparent and compatible to networks globally. Unfortunately these second-generation network standards could not fulfil the dream of having just one set of standards for global networks. The standards in Europe differed from those in Japan and those in America, and so on. Of all the standards, the GSM went the furthest in fulfilling technical and commercial expectations.

But, again, none of the standards in the second generation was able to fulfil the globalisation dream of the standardisation bodies. This would be fulfilled by the third-generation mobile systems. It is expected that these third-generation systems will be predominantly oriented towards data traffic, compared with the second-generation networks that were carrying predominantly voice traffic.

The major standardisation bodies that play an important role in defining the specifications for the mobile technology are:

Fundamentals of Cellular Network Planning & Optimisation A.R. Mishra.
© 2004 John Wiley & Sons, Ltd. ISBN: 0-470-86267-X

- ITU (*International Telecommunication Union*): The ITU, with headquarters in Geneva, Switzerland, is an international organisation within the United Nations, where governments and the private sector coordinate global telecom networks and services. The ITU-T is one of the three sectors of ITU and produces the quality standards covering all the fields of telecommunications.

- ETSI (*European Telecommunication Standard Institute*): This body was primarily responsible for the development of the specifications for the GSM. Owing to the technical and commercial success of the GSM, this body will also play an important role in the development of third-generation mobile systems. ETSI mainly develops the telecommunication standards throughout Europe and beyond.

- ARIB (*Alliance of Radio Industries and Business*): This body is predominant in the Australasian region and is playing an important role in the development of third-generation mobile systems. ARIB basically serves as a standards developing organisation for radio technology.

- ANSI (*American National Standards Institute*): ANSI currently provides a forum for over 270 ANSI-accredited standards developers representing approximately 200 distinct organisations in the private and public sectors. This body has been responsible for the standards development for the American networks.

- 3GPP (*Third Generation Partnership Project*): This body was created to maintain overall control of the specification design and process for third-generation networks. The result of the 3GPP work is a complete set of specifications that will maintain the global nature of the 3G networks.

1.2 MOBILE NETWORK EVOLUTION

Mobile network evolution has been categorised into 'generations' as shown in Figure 1.1. A brief overview on each generation is given below.

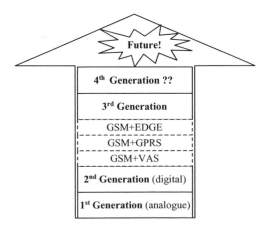

Figure 1.1 Evolution of mobile networks

1.2.1 The First-generation System (Analogue)

The first-generation mobile system started in the 1980s was based on analogue transmission techniques. At that time, there was no worldwide (or even Europe-wide) coordination for the development of technical standards for the system. Nordic countries deployed Nordic Mobile Telephones or NMTs, while UK and Ireland went for Total Access Communication System or TACS, and so on. Roaming was not possible and efficient use of the frequency spectrum was not there.

1.2.2 The Second-generation System (Digital)

In the mid-1980s the European commission started a series of activities to liberalise the communications sector, including mobile communications. This resulted in the creation of ETSI, which inherited all the standardisation activities in Europe. This saw the birth of the first specifications, and the network based on digital technology; it was called the Global System for Mobile Communication or GSM. Since the first networks appeared at the beginning of 1991, GSM has gradually evolved to meet the requirements of data traffic and many more services than the original networks.

- GSM (*Global System for Mobile Communication*): The main elements of this system are the BSS (Base Station Subsystem), in which there are the BTS (Base Transreceiver Station) and BSC (Base Station Controllers); and the NSS (Network Switching Subsystem), in which there is the MSC (Mobile Switching Centre); VLR (Visitor Location Register); HLR (Home Location Register); AC (Authentication Centre), and EIR (Equipment Identity Register) (see Figure 1.4). This network is capable of providing all the basic services such as speech and data services up to 9.6 kbps, fax, etc. This GSM network also has an extension to the fixed telephony networks.

- GSM and VAS (*Value Added Services*): The next advancement in the GSM system was the addition of two platforms, called Voice Mail System (VMS) and the Short Message Service Centre (SMSC). The SMSC proved to be incredibly commercially successful, so much so that in some networks the SMS traffic constitutes a major part of the total traffic. Along with the VAS, IN (INtelligent services) also made its mark in the GSM system, with its advantage of giving the operators the chance to create a whole range of new services. Fraud management and 'pre-paid' services are the result of the IN service.

- GSM and GPRS (*General Packet Radio Services*): As the requirement for sending data on the air-interface increased, new elements such as SGSN (Serving GPRS) and GGSN (Gateway GPRS) were added to the existing GSM system. These elements made it possible to send packet data on the air-interface. This part of the network handling the packet data is also called the 'packet core network'. In addition to the SGSN and GGSN, it also contains the IP routers, firewall servers and DNS (domain name servers). This enables wireless access to the Internet and the bit rate reaching to 150 kbps in optimum conditions.

- GSM and EDGE (*Enhanced Data rates in GSM Environment*): With both voice and data traffic moving on the system, the need was felt to increase the data rate. This was done by using more sophisticated coding methods over the Internet and thus increasing the data rate up to 384 kbps.

1.2.3 Third-generation Networks (WCDMA in UMTS)

In EDGE, high-volume movement of data was possible, but still the packet transfer on the air-interface behaves like a circuit switch call. Thus part of this packet connection efficiency is lost in the circuit switch environment. Moreover, the standards for developing the networks were different for different parts of the world. Hence, it was decided to have a network that provides services independent of the technology platform and whose network design standards are same globally. Thus, 3G was born. In Europe it was called UMTS (Universal Terrestrial Mobile System), which is ETSI-driven. IMT-2000 is the ITU-T name for the third-generation system, while cdma2000 is the name of the American 3G variant. WCDMA is the air-interface technology for the UMTS. The main components include BS (base station) or node B, RNC (radio network controller) apart from WMSC (wideband CDMA mobile switching centre) and SGSN/GGSN. This platform offers many Internet-based services, along with video phoning, imaging, etc.

1.2.4 Fourth-generation Networks (All-IP)

The fundamental reason for the transition to the All-IP is to have a common platform for all the technologies that have been developed so far, and to harmonise with user expectations of the many services to be provided. The fundamental difference between the GSM/3G and All-IP is that the functionality of the RNC and BSC is now distributed to the BTS and a set of servers and gateways. This means that this network will be less expensive and data transfer will be much faster.

1.3 INFORMATION THEORY

1.3.1 Multiple-access Techniques

The basic concept of multiple access is to permit transmitting stations to transmit to receiving stations without any interference. Sending the carriers separated by frequency, time and code can achieve this (Figure 1.2).

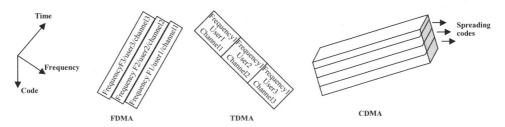

Figure 1.2 Generic multiple-access methods

Frequency-division Multiple Access (FDMA)

This is the most traditional technique in radio communications, relying on separation of the frequencies between the carriers. All that is required is that the transmitters should transmit on the different frequencies and their modulation should not cause the carrier bandwidths to overlap. There should be as many users as possible utilising the frequencies. This multiple access method is used in the first-generation analogue cellular networks. The advantage of the FDMA system is that transmission can be without coordination or synchronisation. The constraint is the limited availability of frequencies.

Time-division Multiple Access (TDMA)

As mobile communications moved on to the second generation, FDMA was not considered an effective way for frequency utilisation, so time-division multiple access was introduced. Thus, as shown in Figure 1.2, many users can use the same frequency as each frequency can be divided into small slots of time called time slots, which are generated continuously.

Code-division Multiple Access (CDMA)

By utilising the spread spectrum technique, code-division multiple access combines modulation and multiple access to achieve a certain degree of information efficiency and protection. Initially developed for military applications, it gradually developed into a system that gave the promise of better bandwidth and service quality in an environment of spectral congestion and interference. In this technology, every user is assigned a separate code/s depending upon the transaction. One user may have several codes in certain conditions. Thus, separation is not based on frequency or time, but on the basis of codes. These codes are nothing but very long sequences of bits having a higher bit rate than the original information. The major advantage of using CDMA is that there is no plan for frequency re-use, the number of channels is greater, there is optimum utilisation of bandwidth, and the confidentiality of information is well protected.

1.3.2 Modulations

Gaussian Minimum Phase-shift Keying (GMSK)

GMSK is the modulation method for signals in GSM. It is a special kind of modulation method derived from minimum phase-shift keying (MSK). Thus it a falls under the frequency modulation scheme. The main disadvantage of MSK is that it has a relatively wide spectrum of operation, so GMSK was chosen to be the modulation method as it utilises the limited frequency resources better. GMSK modulation works with two frequencies and is able to shift easily between the two. The major advantage of GMSK is that it does not contain any amplitude modulation portion, and the required bandwidth of the transmission frequency is 200 kHz, which is an acceptable bandwidth by GSM standards. This is the modulation scheme used in GSM and GPRS networks.

Octagonal Phase-shift Keying (8-PSK)

The reason behind the enhancement of data handling in the 2.5-generation networks such as GPRS/EGPRS is the introduction of octagonal phase-shift keying or 8-PSK. In this scheme, the modulated signal is able to carry 3 bits per modulated symbol over the radio path as compared to 1 bit in the GMSK modulated path. But this increase in data throughput is at the cost of a decrease in the sensitivity of the radio signal. Therefore the highest data rates are provided within a limited coverage. This is the modulation scheme used in the EGPRS/EDGE networks.

Quadrature Phase-shift Keying (QPSK)

To demodulate the frequency modulated signal, phase-shift keying or PSK has been used as the preferred scheme. In PSK, the phase of the transmitted waveform is changed instead of its frequency. In PSK, the number of phase changes is two. A step forward is the assumption that the number of phase changes is more than two (i.e. four), which is the case with quadrature phase-shift keying or QPSK. This enables the carrier to carry four bits instead of two, so effectively doubling the capacity of the carrier. For this reason, QPSK is the modulation scheme chosen in WCDMA Networks.

1.3.3   The OSI Reference Model

The basic idea behind development of the Open System Interconnection (OSI) reference model by the ITU was to separate the various parts that form a communication system. This was possible by layering and modularisation of the functions that were performed by various layers (parts of the communication system). Although initially developed for communications between computers, this model is being extensively used in the telecommunication field, especially mobile communications.

Basic Functions of the OSI Reference Model

Each layer shown in Figure 1.3 communicates with the layer above or below it. No two adjacent layers are dependent, so the lower layer does not worry about the content of the information that it is receiving from the layer above. So, communication between adjacent layers is direct, while communication with the other layers is indirect. Each node has this same reference model. When communicating with layers in another node, each layer

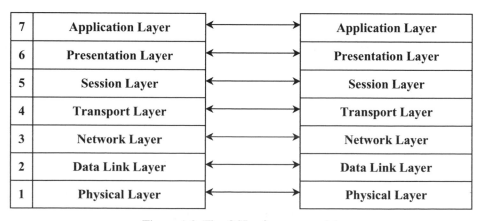

Figure 1.3 The OSI reference model

can communicate with its counterpart in that node (e.g. physical layer with physical layer, transport layer with transport layer). This means that all messages are exchanged at the same level/layer between two network elements, and this is known as a peer-to-peer protocol. All data exchange in a mobile network uses a peer-to-peer protocol.

The Seven Layers of the OSI Reference Model

Layer 1: Physical Layer

The physical layer is so called because of its 'physical' nature. It can be copper wire, an optical fibre cable, radio transmitter or even a satellite connection. This layer is responsible for the actual transmission of data. It transmits the information that it receives from layer 2 without any changes except for the information needed to synchronise with the physical layer of the next node where the information is to be sent.

Layer 2: Data Link Layer

The function of this layer is to pack the data. Data packaging is based on a high-level data link control protocol. This layer combines the data into packets or frames and sends it to the physical layer for transmission. Layer 2 does the error detection and correction and forms an important part in protocol testing, as the information from layer 3 is sent to layer 2 to be framed into packets that can be transferred over layer 1.

Layer 3: Network Layer

This layer is responsible for giving all the information related to the path that a data packet has to take and the final destination it has to reach. Thus, this layer gives the routing information for the data packets.

Layer 4: Transport Layer

This layer is a boundary between the physical elements and logical elements in a network and provides a communication service to the higher layers. Layer 4 checks the consistency of the message by performing end-to-end data control. This layer can perform error detection (but no error correction), and can cater for a reduced flow rate to enable re-transmission of data. Thus, layer 4 provides flow control, error detection and multiplexing of the several transport connections on one network connection.

Layer 5: Session Layer

This layer enables synchronisation between two applications. Both nodes use layer 5 for coordination of the communication between them. This means that it does the application identification but not the management of the application.

Layer 6: Presentation Layer

This layer basically defines and prepares the data before it is sent to the application layer. Layer 6 presents the data to both sides of the network in the same way. This layer is capable of identifying the type of the data, and changes the length by compression or de-compression depending upon the need, before sending it to the application layer.

Layer 7: Application Layer

The application layer itself does not contain any application but acts as an interface between the communication process (layers 1–6) and the application itself. Layer 1 is medium-dependent while layer 7 is application-dependent.

1.4 SECOND-GENERATION MOBILE NETWORKS (GSM)

Of all the second-generation mobile systems, GSM is the most widely used. In this section we will briefly go through its important constituents. It is divided into three major parts (base station subsystem, network subsystem, and network management system) as shown in Figure 1.4.

1.4.1 Base Station Subsystem (BSS)

The BSS consists of the base transreceiver station (BTS), base station controller (BSC) and transcoder sub-multiplexer (TCSM). The latter is sometimes physically located at the MSC. Hence the BSC also has three standardised interfaces to the fixed network, namely A_{bis}, A and X.25.

Base Transceiver Station (BTS)

This manages the interface between the network and the mobile station. Hence, it performs the important function of acting as a hub for the whole of the network infrastructure. Mobile terminals are linked to the BTS through the air-interface. Transmission and reception at the BTS with the mobile is done via omnidirectional or directional antennas (usually having 120-degree sectors). The major functions of the base station are transmission of

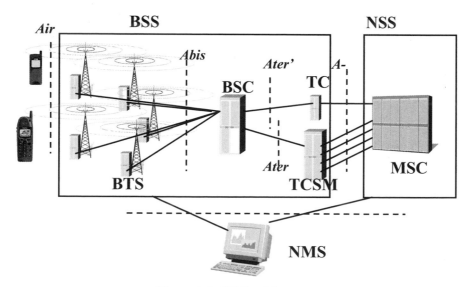

Figure 1.4 GSM architecture

signals in the desired format, coding and decoding of the signals, countering the effects of multi-path transmission by using equalisation algorithms, encryption of the data streams, measurements of quality and received signal power, and operation and management of the base station equipment itself.

Base Station Controller (BSC)

This controls the radio subsystem, especially the base stations. The major functions of the base station controller include management of the radio resources and handover. It is also responsible for control of the power transmitted, and it manages the O&M and its signalling, security configurations and alarms.

1.4.2 Network Subsystem (NSS)

The network subsystem acts as an interface between the GSM network and the public networks, PSTN/ISDN. The main components of the NSS are MSC, HLR, VLR, AUC, and EIR.

Mobile Switching Centre (MSC)

The MSC (or switch as it is generally called) is the single most important element of the NSS as it is responsible for the switching functions that are necessary for interconnections between mobile users and other mobile and fixed network users. For this purpose, MSC makes use of the three major components of the NSS, i.e. HLR, VLR and AUC.

Home Location Register (HLR)

The HLR contains the information related to each mobile subscriber, such as the type of subscription, services that the user can use, the subscriber's current location and the mobile equipment status. The database in the HLR remains intact and unchanged until the termination of the subscription.

Visitor Location Register (VLR)

The VLR comes into action once the subscriber enters the coverage region. Unlike the HLR, the VLR is dynamic in nature and interacts with the HLR when recoding the data of a particular mobile subscriber. When the subscriber moves to another region, the database of the subscriber is also shifted to the VLR of the new region.

Authentication Centre (AUC)

The AUC (or AC) is responsible for policing actions in the network. This has all the data required to protect the network against false subscribers and to protect the calls of regular subscribers. There are two major keys in the GSM standards: the encryption of communications between mobile users, and the authentication of the users. The encryption keys are held both in the mobile equipment and the AUC and the information is protected against unauthorised access.

Equipment Identity Registers (EIR)

Each item of mobile equipment has its own personal identification, which is denoted by a number – the International Mobile Equipment Identity (IMEI). The number is installed during the manufacture of the equipment and states its conformation to the GSM standards. Whenever a call is made, the network checks the identity number; if this number is not found on the approved list of authorised equipment, access is denied. The EIR contains this list of authorised numbers and allows the IMEI to be verified.

1.4.3 Network Management System (NMS)

The main task of the network management system is to ensure that the running of the network is smooth. For this purpose, it has four major tasks to perform: network monitoring, network development, network measurements, and fault management.

Once the network is up and running, the NMS starts to monitor its performance. If it sees a fault it generates the relevant alarm. Some faults may be corrected through the NMS itself (mostly software oriented), while for others site visits are be required.

The NMS is also responsible for the collection of data and the analysis of its performance, thereby leading to accurate decisions related to the optimisation of the network. The capacity and the configuration of the NMS are dependent upon the size (both in terms of capacity and geographical area) and the technological needs of the network.

1.4.4 Interfaces and Signalling in GSM

As can be seen from Figure 1.4, there are some interfaces and signalling involved in the GSM system. Here, we will briefly discuss interfaces and signalling.

Interfaces

Air Interface
The air interface is the central and most important interface in every mobile system. The importance of this interface arises from the fact that it is the only interface the mobile subscriber is exposed to, and the quality of this interface is crucial for the success of the mobile network. The quality of this interface primarily depends upon the efficient usage of the frequency spectrum that is assigned to it.

In the FDMA system, one specific frequency is allocated to one user engaged in a call. When there are numerous calls, the network tends to get overloaded, leading to failure of the system. In a full-rate (FR) system, eight time slots (TS) are mapped on every frequency, while in the half-rate (HR) system, sixteen TSs are mapped on every frequency. In the TDMA system, only impulse-like signals are sent periodically, unlike with FDMA where signals are assigned permanently. Thus, by combining the advantages of both the techniques, TDMA allows seven other channels to be served on the same frequency. As mentioned earlier, GSM uses the gaussian minimum phase-shift keying (GMSK) modulation technique.

For the full-rate system, every impulse on frequency is called a burst. Every burst corresponds to a time slot. So:

1 TDMA frame = 8 bursts

Uplink frequency band = 890–915 MHz

Downlink frequency band = 935–960 MHz

Frequency-division multiplexing divides each of the frequency ranges into 124 channels of 200 kHz width. Since this is required for both transmission and reception, 124 duplex communication channels are produced.

Now, channel bit rate $= D$ per unit time.

As one TDMA frame has eight bursts, hence each burst has a bit rate $d = D/8$.

This means that each time slot (TS) is of duration 577 microseconds (μs). Thus, one frame has duration 4.615 milliseconds (ms).

There are many types of frames in the air-interface, as shown in Table 1.1.

Table 1.1   GSM frame hierarchy

multiframe$_{(51)}$ = 51 TDMA frames
multiframe$_{(26)}$ = 26 TDMA frames
superframe$_{(51)}$ = 51 multiframes
superframe$_{(26)}$ = 26 multiframes
hyperframe = 2028 superframes.

Each frame is of 577 μs and carries a communication channel in which a message element called as packet is transmitted periodically.

The physical channel of the TDMA frames carry logical channels that transport user data and signalling information. Thus, there are traffic channels and signalling channels in the TDMA frame. The traffic channels (TCH) transmit either data or voice signals and can be HR or FR. HR provides a bit rate for coded speech of 6.5 kbps and FR has double that capacity, i.e. 13 kbps. The signalling channels may contain the information such as broadcast channels (e.g. BCCH, SCH, FCH); common control channels (e.g. AGCH, PCH, RACH), dedicated channels (e.g. DCCH, SDCCH), and associated channels (e.g. FACCH, SACCH, SDCCH).

A_{bis} Interface

The A_{bis} interface is the interface between the BTS and the BSC. It is a PCM interface, i.e. it is defined by the 2 Mbps PCM link. Thus it has a transmission rate of 2.048 Mbps, having 32 channels of 64 kbps each. As the traffic channel is 13 kbps on the air interface, while A_{bis} is 64 kbps, hence multiplexing and transcoding do conversion from 64 kbps on the A_{bis} interface to 13 kbps on the air interface. Four channels are multiplexed into one PCM channel while traffic channels are transcoded up to 64 kbps. TCSM are usually located at the MSC to save the transmission costs but can also be located at the BSC site. LAPD signalling, along with TRX management (TRXM), common channel management (CCM), radio link management (RLM) and dedicated channel management (DCM) form a part of this interface.

A Interface

The A interface is present between the TCSM and MSC or physically between the MSC and BSC (generally, TCSM are physically located at the MSC). This interface consists of one or more PCM links each having a capacity of 2048 Mbps. There are two parts of the A-interface – one from the BTS to the TRAU where the transmitted payload is compressed, and one between the TRAU and MSC where all the data is uncompressed. The TRAU is typically located between the MSC and BSC and should be taken into consideration when dealing with this interface. SS7 signalling is present on the A interface.

Signalling

$LAPD_m$

$LAPD_m$ stands for 'modified link access protocol for D-channel'. This is a modified and optimised version of the LAPD signalling for the GSM air interface. The frame structure consists of 23 bytes and is present in three formats: A-, B- and A_{bis}. Both the A- and B-formats are used for both the uplink and downlink, while the A_{bis} is used only for downlinking.

SS7

SS7 provides the basis of all the signalling traffic on all the NSS interfaces. SS7 (signalling system no. 7) is a signalling standard developed by the ITU. This provides the protocols by which the network elements in the mobile (and telephone) networks can exchange information. It is used between the BSC and MSC. It is capable of managing the signalling information in complex networks. It is used for call setup and call management, features such as roaming, authentication, call forwarding, etc. It is not allocated permanently and is required only for the call setup and call release function. It is also known as Common Channel Signalling No. 7 (CCS#7).

X.25

X.25 links the BSC to the O&M centre. It is an ITU-developed signalling protocol, allowing communication between remote devices. This is a packet-switched data network protocol that allows both data and control information flow between the host and the network. By utilizing connection-oriented services, it makes sure that the packets are transferred in order.

1.5 THIRD-GENERATION MOBILE NETWORKS

Third-generation mobile networks are designed for multimedia communication, thereby enhancing image and video quality, and increasing data rates within public and private networks. In the standardization forums, WCDMA technology emerged as the most widely adopted third-generation air-interface. The specification was created by the 3GPP and the name WCDMA is widely used for both the FDD and TDD operations. 3G networks consist of two major parts: the radio access network (RAN) and the core network (CN), as shown in Figure 1.5. The RAN consists of both the radio and transmission parts.

1.5.1 Radio Access Network (RAN)

The main elements in this part of the network are the base station (BS) and the radio network controller (RNC). The major functions include management of the radio resources and telecommunication management.

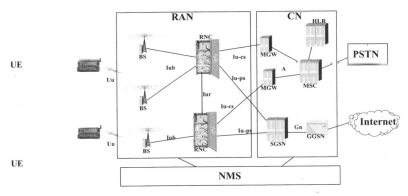

Figure 1.5 Third-generation system (WCDMA)

Base Station (BS)

The base station in 3G is also known as node B. The base station is an important entity as an interface between the network and WCDMA air interface. As in second-generation networks, transmission and reception of signals from the base station is done through omnidirectional or directional antennas. The main functions of the BS include channel coding, interleaving, rate adaptation, spreading, etc., along with processing of the air-interface.

Radio Network Controller (RNC)

This acts as an interface between the base station and the core network. The RNC is responsible for control of the radio resources. Also, unlike in GSM, the RNC in conjunction with the base station is able to handle all the radio resource functions without the involvement of the core network. The major functions of the RNC involve load and congestion control of the cells, admission control and code allocation, routing of the data between the I_{ub} and I_{ur} interfaces etc.

1.5.2 Core Network (CN)

The core network in 3G networks consists of two domains: a circuit-switched (CS) domain and a packet-switched (PS) domain. The CS part handles the real-time traffic and the PS part handles the other traffic. Both these domains are connected to other networks (e.g. CS to the PSTN, and PS to the public IP network). The protocol design of the UE and UTRAN is based on the new WCDMA technology, but the CN definitions have been adopted from the GSM specifications. Major elements of the CN are WMSC/VLR, HLR, MGW (media gateway) on the CS side, and SGSN (serving GPRS support node) and GGSN (gateway GPRS support node) on the PS side.

WCDMA Mobile Switching Centre (WMSC) and VLR (Visitor Location Register)

The switch and database is responsible for call control activities. WMSC is used for the CS transactions, and the VLR function holds information on the subscriber visiting the region, which includes the mobile's location within the region.

Gateway Mobile Switching Centre (GMSC)

This is the interface between the mobile network and the external CS networks. This establishes the call connections that are coming in and going out of the network. It also finds the correct WMSC/VLR for the call path connection.

Home Location Register (HLR)

This is the database that contains all the information related to the mobile user and the kind of services subscribed to. A new database entry is made when a new user is added to the system. It also stores the UE location in the system.

Serving GPRS Support Node (SGSN)

The SGSN maintains an interface between the RAN and the PS domain of the network. This is mainly responsible for mobility management issues like the registration and update of the UE, paging-related activities, and security issues for the PS network.

Gateway GPRS Support Node (GGSN)

This acts as an interface between the 3G network and the external PS networks. Its functions are similar to the GMSC in the CS domain of CN, but for the PS domain.

1.5.3 Network Management System in 3G Networks

As the network technology evolved from 2G to 3G, so did the network management systems. The NMS in 3G systems will be capable of managing packet-switched data also, as against voice- and circuit-switched data in 2G systems. The management systems in 3G need to be more efficient (i.e. more work possible from the NMS rather than visiting the sites). These systems will be able to optimise the system quality in a more efficient way. The management systems are also expected to handle both the multi-technology (i.e. 2G to 3G) and multi-vendor environments.

1.5.4 Interfaces and Signalling in 3G Networks

As can be seen from Figure 1.5, there are some interfaces and signalling involved in 3G systems. Here, we will briefly discuss interfaces and signalling.

Interfaces

Uu/WCDMA Air-interface
Uu or the WCDMA air-interface is the most important interface in 3G networks. The Uu interface works on WCDMA principles wherein all users are assigned one code, which varies with the transaction. As shown in Figure 1.2, each user has a separate spreading code. Unlike with GSM, here every user uses the same frequency band. There are three variants of the CDMA, which are direct-sequence WCDMA frequency-division duplex

(DS-WCDMA-FDD), direct-sequence WCDMA time-division duplex (DS-WCDMA-TDD), and multi-carrier WCDMA (MC-WCDMA). In the initial phase of the WCDMA network rollouts, the FDD variant is used. Initially, 'wideband' was introduced because the Euro-Japanese CDMA version used a wider bandwidth than the American version of CDMA.

Frequency Bands in WCDMA-FDD
The air-interface transmission directions are separated at different frequencies with a duplex distance of 190 MHz (Figure 1.6). The uplink frequency band is 1920–1980 MHz and the downlink is 2110–2170 MHz. There are several bandwidths defined for the WCDMA system (e.g. 5, 10 and 20 MHz), but 5 MHz is the one that is currently being used in network development. Although the bandwidth is 5 MHz, the effective bandwidth will be 3.84 MHz, as the guard band takes up 0.6 MHz from each side (i.e. 1.2 MHz in total makes up the guard band).

I_{ub} Interface
This is the interface that connects the base station to the RNC. This is standardized as an open interface, unlike in the GSM where the interface between the BTS and BSC was not an open one. This consists of common signalling links and traffic termination points. Each of these traffic termination points is controlled by dedicated signalling links. One traffic termination point is capable of controlling more than one cell. I_{ub} interface signalling, known as Node B Application Part (NBAP), has two important constituents. These are Common NBAP (C-NBAP) and Dedicated NBAP (D-NBAP), which are used for common and dedicated signalling links respectively.

I_{ur} Interface
This is a unique interface in WCDMA networks. I_{ur} is an interface between RNC and RNC. There was no such interface in the GSM network (i.e. between BSC and BSC). This interface was designed to support the inter-RNC soft-handover functionality. This interface also supports the mobility between the RNCs, both the common and dedicated channel

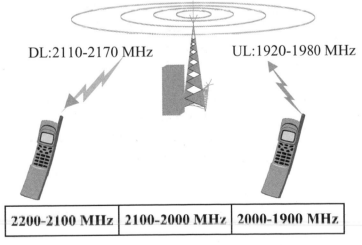

Figure 1.6 Frequency bands for WCDMA

traffic and resource management (e.g. transfer of the cell measurements between the cells). The signalling protocol used for this interface is RNSAP (radio network system application part).

I_u Interface

This interface connects the RAN to the CN. It is an open interface and handles the switching, routing and control functions for both the CS and PS traffic. Thus, the interface that connects the RAN to the CS part of CN part is known as the I_{u-cs} interface, while the one that connects the RAN to the PS part of CN is called the I_{u-ps} interface.

Signalling

Signalling in the third-generation WCDMA networks is in three planes: the transport plane, the control plane and the user plane. This is because a 3G network is understood in three layers that contain data flows. These three layers are transport, control and user plane layers as shown in Figure 1.7.

The transport plane is a means to provide connection between a UE and network (i.e an air-interface). It contains three layers: physical, data and network layers. The physical layer is the WCDMA layer (TDD/FDD), while the data layer is responsible for setting-up/maintaining/deleting the radio link, protection, error corrections etc. The data layer also controls the physical layer. The network layer basically contains functions that are required for the transport control plane.

The control plane contains signalling related to the services that are handled by the network. The I_{ub} interface is maintained by NBAP; in the I_u interface it is RANAP; while in I_{ur} it is RNSAP. For signalling between the application and the destination on the physical layer, user plane signalling is utilised; for example on the U_u interface it is DPDCH.

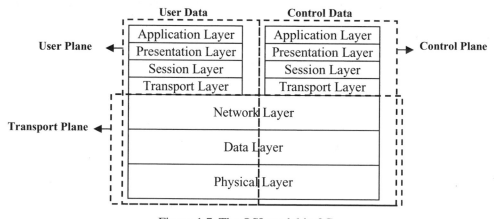

Figure 1.7 The OSI model in 3G

I

Second-generation Network Planning and Optimisation (GSM)

2

Radio Network Planning and Optimisation

Since the early days of GSM development, GSM system network planning has undergone extensive modification so as to fulfill the ever-increasing demand from operators and mobile users with issues related to capacity and coverage. Radio network planning is perhaps the most important part of the whole design process owing to its proximity to mobile users. Before going into details of the process, we first look at some fundamental issues.

2.1 BASICS OF RADIO NETWORK PLANNING

2.1.1 The Scope of Radio Network Planning

The radio network is the part of the network that includes the base station (BTS) and the mobile station (MS) and the interface between them, as shown in Figure 2.1. As this is the part of the network that is directly connected to the mobile user, it assumes considerable importance. The base station has a radio connection with the mobile, and this base station should be capable of communicating with the mobile station within a certain coverage area, and of maintaining call quality standards. The radio network should be able to offer sufficient capacity and coverage.

2.1.2 Cell Shape

In mobile networks we talk in terms of 'cells'. One base station can have many cells. In general, a cell can be defined as the area covered by one sector, i.e. one antenna system. The hexagonal nature of the cell is an artificial shape (Figure 2.1). This is the shape that is closest to being circular, which represents the ideal coverage of the power transmitted

Fundamentals of Cellular Network Planning & Optimisation A.R. Mishra.
© 2004 John Wiley & Sons, Ltd. ISBN: 0-470-86267-X

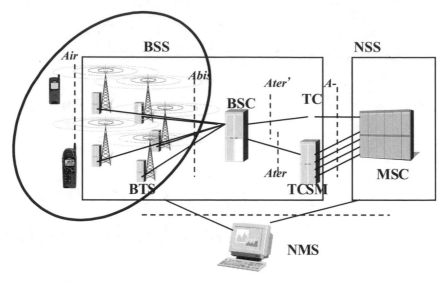

Figure 2.1 The scope of radio network planning

by the base station antenna. The circular shapes are themselves inconvenient as they have overlapping areas of coverage; but, in reality, their shapes look like the one shown in the 'practical' view in Figure 2.2. A practical network will have cells of nongeometric shapes, with some areas not having the required signal strength for various reasons.

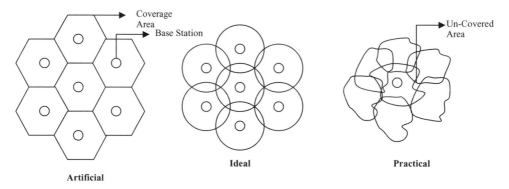

Figure 2.2 Cell shapes

2.1.3 Elements in a Radio Network

Mobile Station (MS)

The mobile station is made up of two parts, as shown in Figure 2.3: the handset and the subscriber identity module (SIM). The SIM is personalised and is unique to the subscriber. The handset or the terminal equipment should have qualities similar to those of fixed phones in terms of quality, apart from being user friendly. The equipment also has functionalities

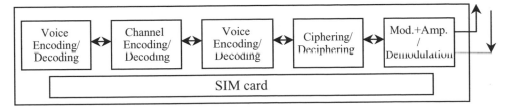

Figure 2.3 Block diagram of a GSM mobile station

like GMSK modulation and demodulation up to channel coding/decoding. It needs to be dual-tone multi-frequency generation and should have a long-lasting battery.

The SIM or SIM card is basically a microchip operating in conjunction with a memory card. The SIM card's major function is to store data for both the operator and subscriber. The SIM card fulfills the needs of the operator and the subscriber as the operator is able to maintain control over the subscription and the subscriber can protect his or her personal information. Thus, the most important SIM functions include authentication, radio transmission security, and storing of the subscriber data.

Base Transceiver Station (BTS)

From the perspective of the radio network-planning engineer the base station is perhaps the most important element in the network as it provides the physical connection to the mobile station through the air interface. And on the other side, it is connected to the BSC via an A$_{bis}$ interface. A simplified block diagram of a base station is shown in Figure 2.4.

The transceiver (TRX) consists basically of a low-frequency unit and a high-frequency unit. The low-frequency unit is responsible for digital signal processing and the high frequency unit is responsible for GMSK modulation and demodulation.

2.1.4 Channel Configuration in GSM

There are two types of channels in the air interface: physical channels and logical channels. The physical channel is all the time slots (TS) of the BTS. There are again two types in this: half-rate (HR) and full-rate (FR). The FR channel is a 13 kbps coded speech or data channel with a raw data rate of 9.6, 4.8 or 2.6 kbps, while the HR supports 7, 4.8 or 2.4 kbps. 'Logical channel' refers to the specific type of information that is carried by the physical channel. Logical channels can also be divided into two types: traffic channels (TCH) and control

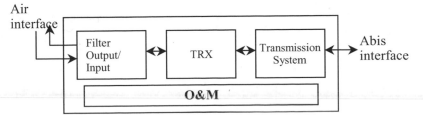

Figure 2.4 Block diagram of a base transceiver station

Table 2.1   Control channels

Channel	Abbreviation	Function/application
Access grant channel (DL)	AGCH	Resource allocation (subscriber access authorisation)
Broadcast common control channel (DL)	BCCH	Dissemination of general information
Cell broadcast channel (DL)	CBCH	Transmits the cell broadcast messages
Fast associated control channel (UL/DL)	FACCH	For user network signalling
Paging channel (DL)	PCH	Paging for a mobile terminal
Random access channel (UL)	RACH	Resource request made by mobile terminal
Slow associated control channel (UL/DL)	SACCH	Used for transport of radio layer parameters
Standalone dedicated control channel (UL/DL)	SDCCH	For user network signalling
Synchronisation channel (DL)	SCH	Synchronisation of mobile terminal

channels (CCH). Traffic channels are used to carry user data (speech/data) while the control channels carry the signalling and control information. The logical control channels are of two types: common and dedicated channels. Table 2.1 summarises the control channel types.

2.2 RADIO NETWORK PLANNING PROCESS

The main aim of radio network planning is to provide a cost-effective solution for the radio network in terms of coverage, capacity and quality. The network planning process and design criteria vary from region to region depending upon the dominating factor, which could be capacity or coverage. The radio network design process itself is not the only process in the whole network design, as it has to work in close coordination with the planning processes of the core and especially the transmission network. But for ease of explanation, a simplified process just for radio network planning is shown in Figure 2.5.

The process of radio network planning starts with collection of the input parameters such as the network requirements of capacity, coverage and quality. These inputs are then used

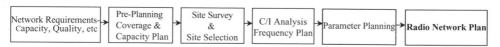

Figure 2.5 The radio network planning process

to make the theoretical coverage and capacity plans. Definition of coverage would include defining the coverage areas, service probability and related signal strength. Definition of capacity would include the subscriber and traffic profile in the region and whole area, availability of the frequency bands, frequency planning methods, and other information such as guard band and frequency band division. The radio planner also needs information on the radio access system and the antenna system performance associated with it.

The pre-planning process results in theoretical coverage and capacity plans. There are coverage-driven areas and capacity-driven areas in a given network region. The average cell capacity requirement per service area is estimated for each phase of network design, to identify the cut-over phase where network design will change from a coverage-driven to a capacity-driven process. While the objective of coverage planning in the coverage-driven areas is to find the minimum number of sites for producing the required coverage, radio planners often have to experiment with both coverage and capacity, as the capacity require-ments may have to increase the number of sites, resulting in a more effective frequency usage and minimal interference.

Candidate sites are then searched for, and one of these is selected based on the inputs from the transmission planning and installation engineers. Civil engineers are also needed to do a feasibility study of constructing the base station at that site.

After site selection, assignment of the frequency channel for each cell is done in a manner that causes minimal interference and maintains the desired quality. Frequency allocation is based on the cell-to-cell channel to interference (C/I) ratio. The frequency plans need to be fine-tuned based on drive test results and network management statistics.

Parameter plans are drawn up for each of the cell sites. There is a parameter set for each cell that is used for network launch and expansion. This set may include cell service area definitions, channel configurations, handover and power control, adjacency definitions, and network-specific parameters.

The final radio plan consists of the coverage plans, capacity estimations, interference plans, power budget calculations, parameter set plans, frequency plans, etc.

2.2.1 Radio Cell and Wave Propagation

Coverage in a cell is dependent upon the area covered by the signal. The distance travelled by the signal is dependent upon radio propagation characteristics in the given area. Radio propagation varies from region to region and should be studied carefully, before predictions for both coverage and capacity are made. The requirement from the radio planners is generally a network design that covers 100% of the area. Fulfilling this requirement is usually impossible, so efforts are made design a network that covers all the regions that may generate traffic and to have 'holes' only in no-traffic zones.

The whole land area is divided into three major classes – urban, suburban and rural – based on human-made structures and natural terrains. The cells (sites) that are constructed in these areas can be classified as outdoor and indoor cells. Outdoor cells can be further classified as macro-cellular, micro-cellular or pico-cellular (see Figure 2.6).

Macro cells

When the base station antennas are placed above the average roof-top level, the cell is a known as a macro-cell. As the antenna height is above the average roof-top level, the area

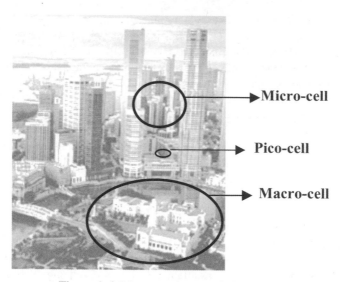

Figure 2.6 Macro-, micro- and pico-cells

that can be covered is wide. A macro-cell range may vary from a couple of kilometres to 35 km, the distance depending upon the type of terrain and the propagation conditions. Hence, this concept is generally used for suburban or rural environments.

Micro-cells
When the base station antennas are below the average roof-top level, then the cell is known as a micro-cell. The area that can be covered is small, so this concept is applied in urban and suburban areas. The range of micro-cells is from a few hundred metres to a couple of kilometres.

Pico-cells
Pico-cells are defined as the same layer as micro-cells and are usually used for indoor coverage.

2.2.2 Wave Propagation Effects and Parameters

The signal that is transmitted from the transmitting antenna (BTS/MS) and received by the receiving antenna (MS/BTS) travels a small and complex path. This signal is exposed to a variety of man-made structures, passes through different types of terrain, and is affected by the combination of propagation environments. All these factors contribute to variation in the signal level, so varying the signal coverage and quality in the network. Before we consider propagation of the radio signal in urban and rural environments, we shall look at some phenomenon associated with the radio wave propagation itself.

Free-space Loss

Any signal that is transmitted by an antenna will suffer attenuation during its journey in free space. The amount of power received at any given point in space will be inversely

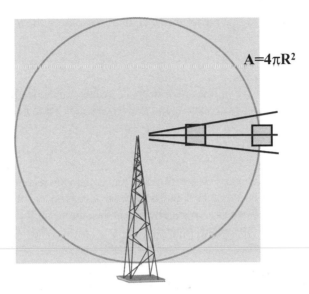

Figure 2.7 Isotropic antenna

proportional to the distance covered by the signal. This can be understood by using the concept of an isotropic antenna. An isotropic antenna is an imaginary antenna that radiates power equally in all directions. As the power is radiated uniformly, we can assume that a 'sphere' of power is formed, as shown in Figure 2.7.

The surface area of this power sphere is:

$$A = 4\pi R^2 \tag{2.1}$$

The power density S at any point at a distance R from the antenna can be expressed as:

$$S = P^*G/A \tag{2.2}$$

where P is the power transmitted by the antenna, and G is the antenna gain. Thus, the received power P_r at a distance R is:

$$P_r = P^*G_t^*G_r^*(\lambda/4\pi R)^2 \tag{2.3}$$

where G_t and G_r are the gain of the transmitting and receiving antennas respectively. On converting this to decibels we have:

$$P_r(\text{dB}) = P(\text{dB}) + G_t(\text{dB}) + G_r(\text{dB}) + 20\log(\lambda/4\pi) - 20\log d. \tag{2.4}$$

Last two terms in equation 2.4 are together called the path loss in free space, or the free-space loss. The first two terms (P and G_t) combined are called the effective isotropic radiated power, or EIRP. Thus:

$$\text{Free-space loss (dB)} = \text{EIRP} + G_r(\text{dB}) - P_r(\text{dB}). \tag{2.5}$$

The free-space loss can then be given as:

$$L_{\text{dB}} = 92.5 + 20\log f + 20\log d \tag{2.6}$$

where f is the frequency in GHz and d is the distance in km.

Equation 2.6 gives the signal power loss that takes place from the transmitting antenna to the receiver antenna.

Radio Wave Propagation Concepts

Propagation of the radio wave in free space depends heavily on the frequency of the signal and obstacles in its path. There are some major effects on signal behaviour, briefly described below.

Reflections and Multipath

The transmitted radio wave nearly never travels in one path to the receiving antenna, which also means that the transmission of the signal between antennas is never line-of-sight (LOS). Thus, the signal received by the receiving antenna is the sum of all the components of the signal transmitted by the transmitting antenna.

Diffraction or Shadowing

Diffraction is a phenomenon that takes place when the radio wave strikes a surface and changes its direction of propagation owing to the inability of the surface to absorb it. The loss due to diffraction depends upon the kind of obstruction in the path. In practice, the mobile antenna is at a much lower height than the base station antenna, and there may be high buildings or hills in the area. Thus, the signal undergoes diffraction in reaching the mobile antenna. This phenomenon is also known as 'shadowing' because the mobile receiver is in the shadow of these structures.

Building and Vehicle Penetration

When the signal strikes the surface of a building, it may be diffracted or absorbed. If it is to some extent absorbed the signal strength is reduced. The amount of absorption is dependent on the type of building and its environment: the amount of solid structure and glass on the outside surface, the propagation characteristics near the building, orientation of the building with respect to the antenna orientation, etc. This is an important consideration in the coverage planning of a radio network.

Vehicle penetration loss is similar, except that the object in this case is a vehicle rather than a building.

Propagation of a Signal Over Water

Propagation over water is a big concern for radio planners. The reason is that the radio signal might create interference with the frequencies of other cells. Moreover, as the water surface is a very good reflector of radio waves, there is a possibility of the signal causing interference to the antenna radiation patterns of other cells.

Propagation of a Signal Over Vegetation (Foliage Loss)

Foliage loss is caused by propagation of the radio signal over vegetation, principally forests. The variation in signal strength depends upon many factors, such as the type of trees, trunks, leaves, branches, their densities, and their heights relative to the antenna heights. Foliage loss depends on the signal frequency and varies according to the season. This loss can be as high at 20 dB in GSM 800 systems.

Fading of the Signal

As the signal travels from the transmitting antenna to the receiving antenna, it loses strength. This may be due to the phenomenon of path loss as explained above, or it may be due to the Rayleigh effect. Rayleigh (or Rician) fading is due to the fast variation of the signal level both in terms of amplitude and phase between the transmitting and receiving antennas when there is no line-of-sight. Rayleigh fading can be divided into two kinds: multipath fading and frequency-selective fading.

Arrival of the same signal from different paths at different times and its combination at the receiver causes the signal to fade. This phenomenon is multipath fading and is a direct result of multipath propagation. Multipath fading can cause fast fluctuations in the signal level. This kind of fading is independent of the downlink or uplink if the bandwidths used are different from each other in both directions.

Frequency-selective fading takes place owing to variation in atmospheric conditions. Atmospheric conditions may cause the signal of a particular frequency to fade. When the mobile station moves from one location to another, the phase relationship between the various components arriving at the mobile antenna changes, thus changing the resultant signal level. Doppler shift in frequency takes place owing to the movement of the mobile with respect to the receiving frequencies.

Interference

The signal at the receiving antenna can be weak by virtue of interference from other signals. These signals may be from the same network or may be due to man-made objects. However, the major cause of interference in a cellular network is the radio resources in the network. There are many radio channels in use in a network that use common shared bandwidth. The solution to the problem is accurate frequency planning, which is dealt with later in the chapter. The mobile station may experience a slow or rapid fluctuation in the signal level in a radio network. This may be due to one or more of the factors discussed above, and as shown in Figure 2.8. These factors form the basis of cell coverage criteria.

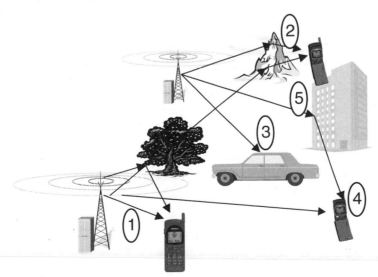

Figure 2.8 Factors affecting wave propagation: (1) direct signal; (2) diffraction; (3) vehicle penetration; (4) interference; (5) building penetration

2.2.3 Dimensioning

The dimensioning exercise is to identify the equipment and the network type (i.e. technology employed) required in order to cater for the coverage and quality requirements, apart from seeing that capacity needs are fulfilled for the next few years (generally 3–5 years). The more accurate the dimensioning is, the more efficient will be network rollout. In practice, network rollout very closely follows the output of network dimensioning/planning. For an efficient network rollout, the equipment has to be ordered well before the planning starts (i.e. after dimensioning), as the equipment orders are placed based on the dimensioning results. Planning engineers should try to do very realistic/accurate dimensioning for each cell site. The inputs that are required for the dimensioning excercise include:

- the geographical area to be covered

- the estimated traffic in each region

- minimum requirements of power in each region and blocking criteria

- path loss

- the frequency band to be used and frequency re-use.

   With the above parameters, the radio planner can predict the number of base stations that will be required for coverage in the specified area to meet the individual quality targets, and to meet the expected increase in traffic in the next few years.

2.3 RADIO NETWORK PRE-PLANNING

Although, in a real scenario, network dimensioning and pre-planning go hand in hand, they have been separated in this chapter for the ease of understanding. Pre-planning can be considered to be the next stage after dimensioning and it is at this stage that some concrete plans related to coverage, capacity and quality are made.

   The major target of the radio planner is to increase the coverage area of a cell and decrease the amount of equipment needed in the network, so obtaining the maximum coverage at minimum cost. Maximum coverage means that the mobile is connected to a given cell at a maximum possible distance. This is possible if there is a minimum signal to noise ratio at both the BTS and MS. Another factor attributing to the path length between the two antennas (BTS and MS) is the propagation loss due to environmental conditions.

Example 1: Calculation of number of sites required in a region

A network is to be designed that should cover an area of $1000\,km^2$.
The base stations to be used are 3-sectored. Each sector (cell) covers a range of 3.0 km
Thus, area covered by each site $= k * R^2$
Where: $k = 1.95$
$\Rightarrow$ Area covered by each site $= 1.95 * 3^2 = 17.55\,km$
Thus: total number of sites $= 1000/17.55 = 56.98 \approx 57$ sites

   Capacity can be understood in simplest terms as the number of mobile subscribers a BTS can cater for at a given time. The greater the capacity, the more mobile subscribers can be connected to the BTS at a given time, thereby reducing the amount of base stations in a given network. This reduction would lead to an increase in the operation efficiency and thereby profits for the network operator. As the number of frequency channels in the GSM is constant (i.e. 125 for GSM 900 in either direction), the re-use of these frequencies determines the number of mobile subscribers who can be connected to a base station. So, efficient frequency planning which includes the assignment of given frequencies and their re-use plays an important part in increasing the capacity of the radio network.

The quality of the network is quite dependent upon the parameter settings. Most of these are implemented during the rollout of the network, just before the launch. In some cases these values are fixed, and in some other cases they are based on measurements done on existing networks. With the first GSM network to be launched in a given region/country, it is helpful for the radio planners to plan these values beforehand for the initial network launch before they have the first measurement results. These may include radio resource management (RRM), mobility management, signalling, handover, and power control parameters. Once there are some measurements available from the initial launch of the network, these parameters then can be fine-tuned. This process becomes a part of the optimisation of the radio network.

2.3.1 Site Survey and Site Selection

When the pre-planning phase is nearing completion, the site search process starts. Based on the coverage plans, the radio planner starts identifying specific areas for prospective sites. There are some points to remember during the process of site selection:

- The process of site selection, from identifying the site to site acquisition, is very long and slow, which may result in a delay of network launch.

- The sites are a long-term investment and usually cost a lot of money.

Therefore, radio planners in conjunction with the transmission planners, installation engineers and civil engineers should try to make this process faster by inspecting the site candidates according to their criteria and coming to a collective decision on whether the candidate site can be used as a cell site or not.

What is a good site for radio planners? A place that does not have high obstacles around it and has a clear view for the main beam can be considered a good radio site. Radio planners should avoid selecting sites at high locations as this may cause problems with uncontrolled interference, apart from giving handover failures.

2.3.2 Result of the Site Survey Process

There are two types of report that are generated in the site survey process. One is at the beginning of the search and the other at the end, which is a report on the site selected. Both reports are very important and should have the desired information clearly given. The site survey request report should stipulate the area where the site candidates should be searched for. The report may contain more specific information such as the primary candidate for search and secondary site candidates – thereby giving the site selection team more specific information on where to put their priorities. Also, this report should contain addresses, maps, and information in the local dialect if possible. The report made after site selection should have more detailed information. This may contain the height of the building/green-field, coordinates, antenna configuration (location, tilt, azimuth, etc.), maps, and a top view of the site with exact location of the base station and the antennas (both radio and transmission).

2.4 RADIO NETWORK DETAILED PLANNING

2.4.1 The Link (or Power) Budget

The detailed radio network plan can be sub-divided into three sub-plans:

(1) link budget calculation,

(2) coverage, capacity planning and spectrum efficiency,

(3) parameter planning.

Link budget calculations give the loss in the signal strength on the path between the mobile station antenna and base station antenna. These calculations help in defining the cell ranges along with the coverage thresholds. Coverage threshold is a downlink power budget that gives the signal strength at the cell edge (border of the cell) for a given location probability. As the link budget calculations basically include the power transmission between the base station (including the RF antenna) and the mobile station antenna, we shall look into the characteristics of these two pieces of equipment from the link budget perspective.

Link budget calculations are done for both the uplink and downlink. As the power transmitted by the mobile station antenna is less than the power transmitted by the base station antenna, the uplink power budget is more critical than the downlink power budget. Thus, the sensitivity of the base station in the uplink direction becomes one of the critical factors as it is related to reception of the power transmitted by the mobile station antenna. In the downlink direction, transmitted power and the gains of the antennas are important parameters. In terms of losses in the equipment, the combiner loss and the cable loss are to be considered. Combiner loss comes only in the downlink calculations while the cable loss has to be incorporated in both directions.

For the other equipment (i.e. the MS), the transmitted power in the uplink direction is very important. To receive the signal transmitted from the BTS antenna even in remote areas, the sensitivity of the MS comes into play. The transmitting and the receiving antenna gains and the cable loss parameters are to be considered on the BTS side.

Important Components of Link Budget Calculations

- MS sensitivity: This factor is dependent upon the receiver noise figure and minimum level of *Eb/No* (i.e. output signal to noise ratio) needed. This is calculated by using the GSM specifications (ETSI GSM recommendation 05.05). The value of MS sensitivity given in these specifications is according to the class of mobile being used. The recommended values of MS sensitivity in GSM 900 and 1800 are −102 dBm and −100 dBm respectively. However, when doing power budget calculations, values given by the manufacturer (or measured values) should be used.

- BTS sensitivity: The sensitivity of the base station is again specified by the ETSI's GSM recommendations 05.05 and is calculated in the same manner as the MS sensitivity. The recommended value of BTS sensitivity is –106 dBm. However, when doing power budget calculations, the value given by the manufacturer (or measured value) should be used.

- Fade margin: This is the difference between the received signal and receiver threshold. Usually a fast fade margin is of importance in power budget calculations. Different values are used for different types of regions, such as 2 dB for dense urban or 1 dB for urban.

- Connector and cable losses: As cables and connectors are used in power transmission, the losses incurred therein should be taken into account. Cable attenuation figures are usually quoted in loss (dB) per 100 m. In such cases, the actual length of the cable should be multiplied by this value to get the theoretical loss taking place in the cable. Sometimes, the theoretical loss may exceed the desired value, so preamplifiers (also known as mast-head amplifiers) may be used to counter the cable loss. Connector losses are usually much less – of the order of 0.1 dB.

- MS and BTS antenna gain: The antennas used for MS and BTS have significantly different gain levels. For obvious reasons, the MS antenna has a lower gain, of the order of 0 dBi, while the BTS antenna gain can vary from 8 dBi to 21 dBi depending upon the type of antenna (omnidirectional versus directional) being used. This gain can be increased by using various techniques, such as antenna diversity (both uplink and downlink).

Example 2: Power budget calculation

Consider a BTS and MS along with the parameters as shown in Figure 2.9.

- RF Power of BST (UL): 42 dBm
- Antenna Gain (Gb): 18 dbm
- Cable Loss (UL/DL): 2 dB
- BTS Sensitivity:- 108 dBm
- Combiner Loss; 2 dB

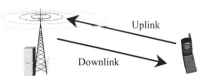

Uplink

Downlink

- RF Power of MS : 32 dBm
- Antenna Gain (Gm): 0 dbm
- Cable Loss (UL/DL): 0 dB
- MS Sensitivity:- 106 dBm

Figure 2.9 Example of a power budget

Uplink calculations

PLu (Path Loss in uplink) = EIRPm (Peak EIRP of Mobile) − Prb (Power Received by the base station)

$$EIRPm = Ptm \text{ (Power transmitted from the MS)} - Losses + Gm$$
$$Losses = Lcm \text{ (cable loss at mobile)} + Lom \text{ (any other loss)}$$
$$Prb = -Gb \text{ (antenna gain)} - Losses + Bs \text{ (BTS sensitivity)}$$
$$Losses = Lcb \text{ (cable loss at BTS)} + Lob \text{ (any other loss)}$$
$$PLu = EIRPm - Prb$$
$$= [Ptm - Lcm - Lom + Gm] - [-Gb + Lcb + Lob + Bs]$$
$$= [32 - 0 + 0 + 0] - [-18 + 2 + 0 + (-108)]$$
$$= 32 + 124 = 156 \, dB$$

Downlink calculations

PLd (Path Loss in downlink) = EIRPb (peak EIRP of BTS) − Prm (Power received by the MS)

$$EIRPb = Ptb\ (Power\ transmitted\ by\ BTS) + Gtb\ (antenna\ gain) − Losses$$
$$Losses = Lcb\ (cable\ loss\ at\ BTS) + Lccb\ (combiner\ loss\ at\ BTS)$$
$$Prm = Ms\ (Mobile\ sensitivity) + Losses − Gm\ (mobile\ antenna\ Gain)$$
$$Losses = Lcm\ (cable\ loss) + Lom\ (any\ other\ loss)$$
$$PLd = EIRPb − Prm$$
$$= [Ptb + Gtb − Lcb − Lccb] − [Ms − Lcm − Lom − Gm]$$
$$= [42 + 18 − 2 − 2] − [−106 − 0 − 0 − 0]$$
$$= 56 + 106 = 162\ dB$$

As can be seen, there is an obvious difference in the results of the uplink and downlink power budget calculations, where the downlink path loss exceeds the uplink power loss. This is an indication that the area covered by the base station antenna radiations is more than the area covered by the mobile station antenna, thereby giving more coverage in the downlink direction. Reducing the power in the downlink direction can reduce this difference but results in a loss of coverage. Another way is to introduce diversity at the BTS, or even to introduce low-noise amplifiers (LNA) at the BTS. Both measures will have a positive impact on the BTS receiver power level. Another power budget calculation for GSM 900 and 1800 system using different classes of mobiles (A and D) is shown in Example 3.

How can there be an improvement in the power budget results? As seen above, apart from varying the power transmitted from the BTS antenna, these results can be improved by using enhanced planning techniques such as frequency hopping and/or by using some enhancements such as receiver diversity, LNA for the uplink directions, and boosters or filters for the downlink directions.

Example 3: Simple power budget calculations

RADIO LINK POWER BUDGET			**MS CLASS**	**1**
GENERAL INFO				
Frequency (MHz):	1800		System:	GSM
RECEIVING END:		**BS**	**MS**	
RX RF-input sensitivity	dBm	-104.00	-100.00	A
Interference degrad. margin	dB	3.00	3.00	B
Cable loss + connector	dB	2.00	0.00	C
Rx antenna gain	dBi	18.00	0.00	D
Diversity gain	dB	5.00	0.00	E
Isotropic power	dBm	-122.00	-97.00	F=A+B+C-D-E
Field strength	dB V/m	7.24	32.24	G=F+Z*
				* Z = 77.2 + 20*log(freq(MHz))
TRANSMITTING END:		**MS**	**BS**	
TX RF output peak power	W	1.00	15.85	
(mean power over RF cycle)	dBm	30.00	42.00	K
Isolator + combiner + filter	dB	0.00	3.00	L
RF-peak power, combiner output	dBm	30.00	33.00	M=K-L
Cable loss + connector	dB	0.00	2.00	N
TX-antenna gain	dBi	0.00	18.00	O
Peak EIRP	W	1.00	79.43	
(EIRP = ERP + 2dB)	dBm	30.00	49.00	P=M-N+O
Path loss due to ant./body loss	dBi	6.00	6.00	Q
Isotropic path loss	dB	146.00	146.00	R=P-F-Q

RADIO LINK POWER BUDGET			**MS CLASS**	**4**
GENERAL INFO				
Frequency (MHz):	900		System:	GSM
RECEIVING END:		**BS**	**MS**	
RX RF-input sensitivity	dBm	-104.00	-102.00	A
Interference degrad. margin	dB	3.00	3.00	B
Cable loss + connector	dB	4.00	0.00	C
Rx antenna gain	dBi	12.00	0.00	D
Isotropic power	dBm	-109.00	-99.00	E=A+B+C-D
Field strength	dB V/m	20.24	30.24	F=E+Z*
				* Z = 77.2 + 20*log(freq(MHz))
TRANSMITTING END:		**MS**	**BS**	
TX RF output peak power	W	2.00	6.00	
(mean power over RF cycle)	dBm	33.00	38.00	K
Isolator + combiner + filter	dB	0.00	3.00	L
RF-peak power, combiner output	dBm	33.00	26.00	M=K-L
Cable loss + connector	dB	0.00	4.00	N
TX-antenna gain	dBi	0.00	12.00	O
Peak EIRP	W	2.00	20.00	
(EIRP = ERP + 2dB)	dBm	33.00	34.00	P=M-N+O
Path loss due to ant./body loss	dBi	9.00	9.00	Q
Isotropic path loss	dB	133.00	133.00	R=P-F-Q

Output and Effect of Link Budget Calculations

- Path loss and received power: This is the main output of the link budget calculations. The losses in signal strength that occur during transmission from the TX antenna to the RX antenna are given by the path loss, while the received power is the result of the path loss phenomenon. All the factors that contribute to increases (e.g. antenna gains) and decreases (e.g. losses due to propagation) are taken into account during the calculations. The better the input data accuracy, the more accurate the results.

- Cell range: If the path loss is lessened, the signal from the transmitter (BTS) antenna will cover more distance, so increasing the area covered by one BTS. Thus, the power budget calculations play a direct role in determining the covered area, and so deciding on the number of base stations that will be required in a network.

- Coverage threshold: The downlink signal strength at the cell border for a given location probability is known the coverage threshold. Although slow fade margin and MS isotropic power can be used to calculate this value, power budget calculations are used for this purpose. Propagation models are used for more accurate calculation of the cell range and coverage area (refer to Example 3).

2.4.2 Frequency Hopping

Frequency hopping (FH) is a technique that basically improves the channel to interference (C/I) ratio by utilising many frequency channels. Employment of the FH technique also improves the link budget due to its effects: frequency diversity and interference diversity.

The frequency diversity technique increases the decorrelation between the various frequency bursts reaching the moving MS. The effects of fading due to propagation conditions reduces, thereby improving the signal level. There are again two types of frequency diversity technique: random FH and sequential FH. Sequential FH is used more in practical network planning as it gives more improvement to the network quality.

If the number of frequency channels increases in the radio network, the number of frequencies used increases in the network, so reducing the interference effect at the mobile station. This leads to an increase in signal level, and an improvement in the power budget.

2.4.3 Equipment Enhancements

Receiver Diversity

Diversity is the most common way to improve the reception power of the receiving antenna. Major diversity techniques are space diversity, frequency diversity, and polarisation diversity. Frequency diversity is also known as frequency hopping.

Space diversity involves installing another antenna at the base station. This means that there are two antennas receiving the signal at the base station instead of one and are separated in *space* by some distance. There is no fixed distance of separation between the antennas, which depends upon the propagation environment. Depending on the environmental conditions, the distance between the main and the diversity antenna can vary from 1 to 15 wavelengths.

Polarisation diversity means that the signals are received using two polarisations that are orthogonal to each other. It can be either vertical–horizontal polarisation or it can be ±45-degree slated polarisation.

Low-noise Amplifiers (LNA)

Where the received power is limited by the use of long cables, low-noise amplifiers can be used to boost the link budget results. As the name suggest, a LNA has a low noise value and can amplify a signal. The LNA is placed at the receiving end. When space diversity is being used, the LNAs should be used on both the main and the diversity antennas, thereby improving the diversity reception. As stated above, this is used for improvement of the uplink power budget.

Power Boosters

Power in the downlink direction can be increased by the use of power amplifiers and power boosters. If the losses are reduced before the transmission by the use of amplifiers, which in turn increases the power, then the configuration is called a power amplifier. However, when the transmission power is increased, then it is done by using the booster. Power amplifiers are located near the transmission antennas while the boosters are located near the base station as shown in Figure 2.10.

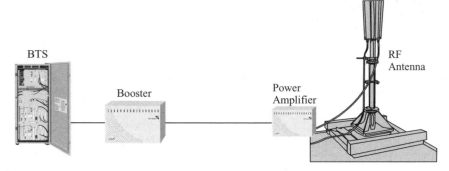

Figure 2.10 Power budget enhancement using a booster and power amplifier

2.4.4 Cell and Network Coverage

The cell and network coverage depend mainly on natural factors such as geographical aspect/propagation conditions, and on human factors such as the landscape (urban, suburban, rural), subscriber behaviour etc. The ultimate quality of the coverage in the mobile network is measured in terms of *location probability*. For that, the radio propagation conditions have to be predicted as accurately as possible for the region.

There are two ways in which radio planners can use propagation models. They can either create their own propagation models for different areas in a cellular network, or they can use the existing standard models, which are generic in nature and are used for a whole area. The advantage of using their own model is that it will be more accurate, but it will also be immensely time-consuming to construct. Usage of the standard models is economical from the time and money perspective, but these models have limited accuracy. Of course, there is a middle way out: the use of multiple generic models for urban, suburban and rural environments in terms of macro-cell or micro-cell structure.

A Macro-cell Propagation Model

The Okumara–Hata model is the most commonly used model for macro-cell coverage planning. It is used for the frequency ranges 150–1000 MHz and 1500–2000 MHz. The range of calculation is from 1 to 20 km. The loss between the transmitting and receiving stations is given as:

$$L = A + B\log f - 13.82\log h_{bts} - a(h_m)(44.9 - 6.55\log h_b)\log d + L_{other} \qquad (2.7)$$

where f is the frequency (MHz), h is the BTS antenna height (m), $a(h)$ is a function of the MS antenna height, d is the distance between the BS and MS (km), L_{other} is the attenuation due to land usage classes, and $a(h_m)$ is given by:

$$a(h_m) = (1.1\log f_c - 0.7)h_m - (1.56\log f_c - 0.8).$$

For a small or medium-sized city:

$$a(h_m) = 8.25(\log 1.54 h_m)^2 - 1.1, \quad \text{for} \quad f_c \leq 200\,\text{MHz} \qquad (2.8)$$

For a large city:

$$a(h_m) = 3.2(\log 11.75 h_m)^2 - 4.97, \quad \text{for} \quad f_c \geq 400\,\text{MHz} \qquad (2.9)$$

The value of the constants A and B varies with frequencies as shown below:

$$A = 69.55 \quad \text{and} \quad B = 26.16 \text{ for } 150{-}1000\,\text{MHz}$$
$$A = 46.3 \quad \text{and} \quad B = 33.9 \text{ for } 1000{-}2000\,\text{MHz}.$$

The attenuation will vary with the type of terrain. This may include losses in an urban environment where small cells are predominant. Then there are foliage losses when forests are present in the landscape. Similarly, the effects of other natural aspects such as water bodies, hills, mountains, glaciers, etc., and the change in behaviour in different seasons have to be taken into account.

A Micro-cell Propagation Model

The most commonly used micro-cellular propagation model is the Walfish–Ikegami model. This is basically used for micro-cells in urban environments. It can be used for the frequency range 800–2000 MHz, for heights up to 50 m (i.e. the height of building + height of the BTS antenna) for a distance of up to 5 km. This model talks about two conditions: line-of-sight (LOS) and no-line-of-sight (NLOS). The path loss formula for the LOS condition is:

$$P = 42.6 + 26 \log d + 20\log f. \qquad (2.10)$$

For the NLOS condition, the path loss is given as:

$$P = 32.4 + 20 \log f + 20 \log d + L_{rds} + L_{ms}. \qquad (2.11)$$

The parameters in the equations above for the model can be understood from Figure 2.11. The values of the rooftop-to-street diffraction loss are dependent upon the street orientation, street width and the frequency of operation. The multi-screen diffraction losses are dependent upon the distance and frequency.

Note: Walfish–Ikegami model can be used also for macro-cells. However, some radio planning engineers do use other models – such as ray tracing – for the micro-cellular environment.

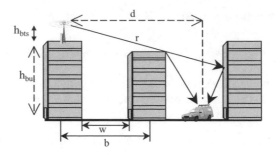

Figure 2.11 W–I model:

d: distance in km

f: frequency in MHz

Lrds: rooftop-street diffraction and scatter loss

Lms: multi-screen diffraction loss

w: road width

b: distance between the centres of two buildings

H_{bu}: height of the building

Application of Propagation Models

The propagation models are usually not applied directly. The reason is that these models were developed taking particular cities into account, and every city has its own characteristics. Changes made to the propagation models are called correction factors and they are based on drive tests results. If there is no existing cellular network, the radio planning engineers install an omni-antenna at a location which would cover all or most of the types of region – dense urban, urban, rural, etc. A drive test is performed and correction factors for the propagation models are thereby determined. One such table of correction factors is shown in Figure 2.11(a).

Category	Offset(dB)	Code
No data	0.00	0
Water	-17.00	1
Open	-17.00	2
Evergreen	-17.00	3
Deciduous	-10.00	4
Low Density Residential	-15.00	5
High Density Residential	-10.00	6
Commercial/ Industrial	-15.00	7
Urban	-3.20	8

Figure 2.11(a) Correction factors

Planned Coverage Area

Based on propagation models, drive tests and correction factors, prediction of coverage areas is done. The sites are located according to the requirements of the network, and the coverage predictions are done as shown in Figure 2.11(b). Usually some radio network planning tools are used for such an exercise.

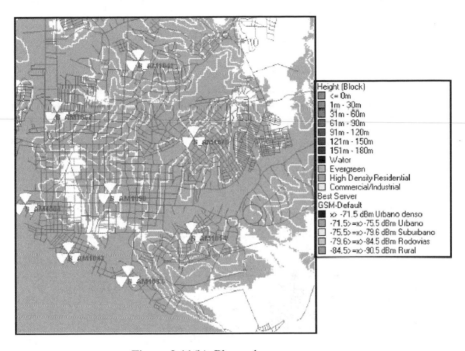

Figure 2.11(b) Planned coverage area

Location Probability

As mentioned above, the quality of coverage is defined in terms of location probability. Location probability can be defined as the probability of the field strength being above the sensitivity level in the target area. For practical purposes, it is considered that a location probability of 50% is equal to the sensitivity of the receiver in the given region. As the received power at the receiver should be higher than the sensitivity, the location probability should therefore be higher than 50%. Earlier we looked at the reasons behind fluctuations and fading of the signal strength. These fluctuations may be more or less than the sensitivity of the receiver. Hence, the design of the radio network incorporates a term known as the fade margin. Planning is done in such as way that the field strength of the signal is higher than the sensitivity by this margin. So, when the fading is taking place (slow fading or shadowing being the most prominent), then the signal level after fading is way above the sensitivity of the receiver.

A more accurate link budget calculation taking into account the propagation model effects is shown in Example 4.

Example 4: Detail Radio Link Power Budget

RADIO LINK POWER BUDGET					MS CLASS:	2
GENERAL INFO						
Frequency (MHz):		900		System:		GSM
RECEIVING END:				BS	MS	
RX RF-input sensitivity		dBm		-106.00	-103.00	A
Interference degrad. margin		dB		2.00	2.00	B
Cable loss + connector		dB		2.50	10.00	C
Rx antenna gain		dBi		16.00	0.00	D
Diversity gain		dB		3.50	0.00	E
Isotropic power		dBm		-121.00	-91.00	F=A+B+C-D-E
Field strength		dBµV/m		8.24	38.24	G=F+Z*
						*$Z = 77.2 + 20*log(freq[MHz])$
TRANSMITTING END:				MS	BS	
TX RF output peak power		W		3.00	33.66	
(mean power over RF cycle)		dBm		34.77	45.27	K
Isolator + combiner + filter		dB		0.00	4.00	L
RF-peak power, combiner output		dBm		34.77	41.27	M=K-L
Cable loss + connector		dB		10.00	2.50	N
TX-antenna gain		dBi		0.00	16.00	O
Peak EIRP		W		0.30	300.00	
(EIRP = ERP + 2dB)		dBm		24.77	54.77	P=M-N+O
Isotropic path loss		dB		145.77	145.77	Q=P-F
CELL SIZES						
COMMON INFO				Region 1	Region 2	GENERAL
MS antenna height (m):				1.5	1.5	1.5
BS antenna height (m):				30.0	30.0	30.0
Standard Deviation (dB):				7.0	7.0	7.0
BPL Average (dB):				25.0	20.0	15.0
BPL Deviation (dB):				7.0	7.0	7.0
OKUMURA-HATA (OH)				Region 1	Region 2	GENERAL
Area Type Correction (dB)				0.0	-4.0	-6.0
WALFISH-IKEGAMI (WI)				Region 1	Region 2	GENERAL
Roads width (m):				30.0	30.0	30.0
Road orientation angle (degrees):				90.0	90.0	90.0
Building separation (m):				40.0	40.0	40.0
Buildings average height (m):				30.0	30.0	30.0
INDOOR COVERAGE				Region 1	Region 2	GENERAL
Propagation Model				OH	OH	OH
Slow Fading Margin + BPL (dB):				32.4	27.4	22.4
Coverage Threshold (dBµV/m):				70.6	65.6	60.6
Coverage Threshold (dBm):				-58.6	-63.6	-68.6
Location Probability over Cell Area(L%):				95.0%	95.0%	95.0%
Cell Range (km):				0.78	1.40	2.22
OUTDOOR COVERAGE						GENERAL
Propagation Model						OH
Slow Fading Margin (dB):						7.4
Coverage Threshold (dBµV/m):						45.6
Coverage Threshold (dBm):						
Location Probability over Cell Area(L%):						95.0%
Cell Range (km):						5.92

2.4.5 Capacity Planning

Capacity planning is a very important process in the network rollout as it defines the number of base stations required and their respective capacities. Capacity plans are made in the pre-planning phase for initial estimations, as well as later in a detailed manner.

The number of base stations required in an area comes from the coverage planning, and the number of transceivers required is derived from capacity planning as it is directly associated with the frequency re-use factor. The frequency re-use factor is defined as the number of base stations that can be implemented before the frequency can be re-used. An example of frequency re-use is shown in Figure 2.12. The maximum number of frequencies in a GSM 900 system is 125 in both the uplink and downlink directions. Each of these frequencies is called a channel. This means that there are 125 channels available in both directions. The minimum frequency re-use factor calculation is based on the C/I ratio. As soon as the C/I ratio decreases, the signal strength starts deteriorating, thereby reducing the frequency re-use factor.

Another factor to keep in mind is the antenna height at the base station. If the antenna height is too high then the signal has to travel a greater distance, so the probability that the signal causes interference becomes greater. The average antenna height should be such that the number of base stations (fully utilised in terms of their individual capacities) is enough for the needed capacity of the network. Of course, as seen above, this depends heavily on the frequency re-use factor.

There are three essential parameters required for capacity planning: estimated traffic, average antenna height, and frequency usage.

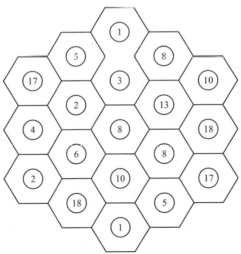

Figure 2.12 Example of frequency re-use

Traffic Estimates

Traffic estimation or modelling is based on theoretical estimates or assumptions, and on studies of existing networks (i.e. experience). Traffic in the network is dependent on the user communication rate and user movement in the network. The user communication rate means how much traffic is generated by the subscriber and for how long. The user movement is an estimate of the user's use of the network in static mode and dynamic mode.

Traffic estimation in the network is given in terms of 'erlangs'. One erlang (1 Erl) is defined as the amount of traffic generated by the user when he or she uses one traffic channel for one hour (this one hour is usually the busy hour of the network). Another term that is frequently used in network planning is 'blocking'. Blocking describes the situation when a user is trying to make a call and is not able to reach a dialled subscriber owing to lack of resources.

Generally, it is assumed that a user will generate about 25 mErl of traffic during the busy hour, and that the average speaking (network usage) will be about 120 seconds. These figures may vary from network to network; some networks use average figures of 35 mErl and 90 s.

Another factor is the user's behaviour in terms of mobility. In the initial years of the GSM, the ratio of static users to dynamic users was almost 0.7, but with rapid changes in technology this ratio may soon become 1.0! User mobility affects handover rates, which in turn affects network capacity planning.

The actual traffic flowing in the network can be calculated by using tables that use the maximum traffic at a base station and the blocking rate. Commonly used Erlang tables are Erlang B and Erlang C. Erlang B assumes that if calls cannot go through then they get dropped (i.e. no queuing possible). Erlang C considers that if a call does not get through then it will wait in a queue. These Erlang tables are good enough for circuit-switched traffic but not for packet switching. We look into the packet switching aspect in later chapters. Erlang B tables are given in Appendix E.

It is important for the radio planner to know the capacity that can be offered by the base station equipment, which means that the traffic handling capability of the transceivers takes precedence for capacity planning. Due to modulation, the modulated stream of bits is sent in *bursts* having a finite duration. These bursts are generally called *time slots* (TS) in GSM; they have a relatively fixed place in the stream and occur after 0.577 ms. The slots have a width of about 200 kHz. In the GSM system this is known as *one time slot*. Due to the modulation schemes (i.e. TDMA), there are eight TS at each frequency in each direction. All these eight time slots can be used for sending the traffic or the signalling information. Channel organization within the time slot should be done in such a way that every time a burst is transmitted, it is utilised completely. A typical time-slot composition is shown in Figure 2.13.

Signalling requires one time slot (e.g. TS0), and the remaining seven time slots can be used for traffic. In this configuration, the number of subscribers who can talk simultaneously is seven, on separate traffic channels (TCH). Now, when the number of transceivers increases in the cell, the traffic and signalling channel allocation also change. Generally,

Figure 2.13 Time slot configuration for single TRX

two transceivers (TRX) would have 15 TCH and one SCH (signalling channel). Four TRX would have 30 TCH and two SCH, which means that the traffic channels increase by $7 + 8 + 7 + 8 \cdots$ for every single increase in the number of transceivers, while the signalling channels will have an increment at every alternate TRX addition (which means that a decrease in SCHs in the TRX increases the TCHs in it).

Average Antenna Height

The concept of the average antenna height is the basis of the frequency re-use pattern determining capacity calculations in a cellular network. The average antenna height is the basis of the cellular environment (i.e. whether it is macro-cellular or micro-cellular). If the average antenna height is low, then the covered area is small in an urban environment. This will lead to the creation of more cells, and hence increase the number of times the same frequency can be re-allocated. Exactly the opposite is the case in a macro-cellular environment. Here the coverage area would be more, so the same frequency can be re-allocated fewer times. All these calculations are based on the interference analysis of the system as well as the topography and propagation conditions.

Frequency Usage and Re-use

Frequency usage is an important concept related to both coverage and capacity usage. Frequency re-use basically means how often a frequency can be re-used in the network. If the average number of the transceivers and the total number of frequencies are known, the frequency re-use factor can be calculated.

Example 5: Frequency re-use factor

If there are 3 TRX that are used per base station and the total number of frequencies available is 27, then the total number of frequencies available for re-use is $27/3 = 9$.

2.4.6 Spectrum Efficiency and Frequency Planning

Spectrum efficiency is simply the maximum utilisation of the available frequencies in a network. In the radio planning process, this is known as frequency planning. Capacity and frequency planning do of course go hand-in-hand, but the concepts described so far in this chapter provide the inputs for frequency planning. A good frequency plan ensures that frequency channels are used in such a way that the capacity and coverage criteria are met without any interference. This is because the total capacity in a radio network in terms of the number of sites is dependent upon two factors: transmission power and interference. The re-use of the BCCH TRX (which contains the signalling time slots) should be greater than that of the TCHs, since it should be the most interference-free.

2.4.7 Power Control

The power that is transmitted both from the mobile equipment and from the base station has a far-reaching effect on efficient usage of the spectrum. Power control is an essential feature in mobile networks, in both the uplink and downlink directions. When a mobile transmits high power, there is enough fade margin in the critical uplink direction, but it can cause interference to other subscriber connections. The power should be kept to a level that the signal is received by the base station antenna above the required threshold without causing interference to other mobiles. Mobile stations thus have a feature such that their power of transmission can be controlled. This feature is generally controlled by the BSS. This control is based on an algorithm that computes the power received by the base station and, based on its assessment, it increases or decreases the power transmitted by the mobile station.

2.4.8 Handover

Handover is the automatic transfer of the subscriber from one cell to another during the call process, without causing any hindrance to the call. There are two main aspects to this: the necessity to find a dedicated mode in the next cell as the mobile is on call, and the switching process being fast enough so as not to drop that call.

So, how does the handover actually take place? There are many processes that can be used, but the one most used is based on power measurements. When a mobile is at the interface of two cells, the BSS measures the power that is received by the base stations of the two cells, and then the one that satisfies the criteria of enough power and least interference is selected. This kind of handover being directly related to power control, it provides an opportunity to improve the efficiency of use of the spectrum.

2.4.9 Discontinuous Transmission

Discontinuous transmission (DTX) is a feature that controls the power of the transmission when the mobile is in 'silent' mode. When the subscriber is not speaking on the mobile, a voice detector in the equipment detects this and sends a burst of transmission bits to BSS, indicating this inactivity. This function of the mobile is called voice-activity detection (VAD). On receiving this stream of bits indicating DTX, the BSS asks the mobile to reduce its power for that period of time, thereby reducing interference in the network and improving the efficiency of the network.

2.4.10 Frequency Hopping

Before we go into the concept and process of frequency hopping, let us understand the frequency assignment criteria in the GSM network. In GSM 900, the frequency bands used are 890–915 MHz in the uplink direction and 935–960 MHz in the downlink direction, which means a bandwidth of 25 MHz in each direction. The whole or some fraction of this band is available to the network operator. The central frequencies start at 200 kHz from the 'edge' of the band and are spread evenly in it. There are 125 frequency slots in this band. The major interference problem is between the adjacent bands because of frequency overlapping at the borders of the individual channels. For this simple reason, the adjacent (and same-frequency) channels) are not used on the cells belonging to the same site.

Frequency hopping is a technique by which the frequency of the signal is changed with every burst in such a way that there is minimum interference in the network, and at the same time allocated channels are used effectively. This process in GSM is also known as slow frequency hopping (SFH). By using SFH, improvement takes place by virtue of *frequency diversity* and *interference diversity*.

- Frequency diversity: Since every burst has a different frequency, it will fade in a different way and time. Thus the decorrelation between each burst increases, thereby increasing the efficiency of the coding signal. The assignment of the frequency can be done by two ways: sequentially or randomly. In the former, the system follows a strict pattern of frequency assignment to each burst; in the latter, it assigns frequencies randomly.

- Interference diversity: If each mobile has one constant frequency, some mobiles may be affected by interference more than others. With the use of frequency hopping, the interference spreads within the system because the interfering signal's effect gets reduced. As the interference becomes less, the frequency spectrum can be utilised better, and hence the capacity of the system increases.

Frequency hopping is of two types: base-band FH and RF FH. In base-band FH, the calls are *hopped* between different TRXs. The number of frequencies used for hopping is correlated to the number of TRXs and is thus constant. In RF FH, the call stays on one TRX but a frequency change takes place with every frame. These frequencies are not included in the hopping sequence, thus effectively creating two layers in each cell, one FH and one non-FH. As RF FH is not correlated to the number of TRXs, it is considered to be more robust and hence used in network deployment frequency planning.

2.4.11 Parameter Planning

The parameters used in a radio network are of two types: fixed and measured. These parameters include those related to signalling, radio resource management (RRM), power control, neighbour cells, etc.

Signalling

Any flow of data in a network requires some additional information that helps the data to reach the destination in the desired fashion. This additional information is known as signalling.

Signalling in GSM is required at all the interfaces, but radio network planners deal mostly with the signalling between the mobile station and base station (shown in Table 2.1). Signalling on all the interfaces except for the air-interface is done at 64 kbps. On the air-interface the signalling can be done either by using the slow associated control channels (SAACH), or by using the main channel itself wherein the signalling channel is sent instead of sending the data – this is known as fast associated control channel (FAACH) signalling. SAACH signalling is 'slow' and hence carries non-urgent messages, such as information containing handover measurement data. FAACH signalling carries information that is more urgent, such as decisions leading to handover of the mobile. Signalling is also required at the air-interface for sending information about the mobile itself even when it is not on a call. Thus, signalling can be in the dedicated phase (i.e. when TCHs have been allocated to the mobile) and in the non-dedicated or idle mode when the mobile is not on a call but is camped on the network.

When the mobile firsts tries to get connected to the network, it requires the help of two channels: the frequency correction channel (FCCH) and the synchronisation channel (SCH). These channels help the mobile to get synchronised and connected to the network. Each mobile, once connected, keeps on receiving information from the base station, and this is done through the broadcast channel (BCCH). Once a call is initiated, the paging channel (PCH) helps in transfer of information indicating that a dedicated channel will be allocated to the mobile. This allocation information comes via the allocation grant channel (AGH). If the mobile needs to send information to the base station, the request is made through the random access channel (RACH). All these channels except RACH are downlink channels.

The above-mentioned channels are logical channels. There are three kinds of physical channel also: the traffic channels for full-rate, half-rate and one-eighth rate (also known as TCH/F, TCH/H and TCH/8 respectively). TCH/F transmits the speech code at 13 kbps; TCH/H transmits speech code at 7 kbps. Although TCH/8 is a traffic channel, its rate is very low at almost one-eighth that of the TCH/F; thus its usage has been limited to signalling.

Radio Resource and Mobility Management

The management of radio resources, functions related to mobile location update, communication management issues such as handover and roaming procedure handling, come under radio resource management (RRM). For these management functions to happen, information flow (traffic and signalling) takes place via three protocols, known as *link protocols* (see Figure 2.14). LAPDm is present over the MS–BTS connection and LAPD over the BTS–BSC connection. MTP (message transfer protocol) is used for signalling transport over the SS7 network.

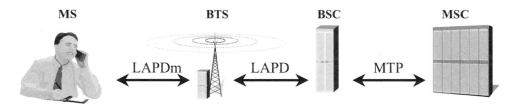

Figure 2.14 Link protocol in a GSM network

RRM procedures basically relate to the processes taking place during transitions between different states of the mobile station, such as the idle state, the dedicated state, during mobility, during handover, when calls are being re-established, etc.

In GSM, there are two states that are defined for a mobile: idle and dedicated. Thus, from an idle state, the mobile station enters a BTS access mode wherein 'access' is granted to the mobile based on whether or not the mobile station is allowed to 'use' the base station. Then the mobile station enters the dedicated mode and starts using the resources until it enters the release mode, i.e. the call ends.

Once a mobile is logged into the network, the transition procedure request always comes from the mobile, and this is done through the random access channel (RACH). As the timing of requests coming from mobiles cannot be predetermined, this may cause problems such as 'collision of call request'. Then factors such as congestion, the call-repetition process, traffic increase, etc., come into the picture. The radio planner should keep the throughput under control by means of parameters such as the number of times the request for a channel can be sent, and the timing of re-requests (which is usually kept 'random').

Paging is another function whereby, when a request from the mobile reaches the MSC, then the MSC 'pages' the requested subscriber information to all the BSCs within a location area. The BSC in turn sends this information to the BTS to find the subscriber through the paging and access grant channel (PAGCH). The tasks assigned to the BSC and BTS may vary from network to network.

The subscriber should be able to connect to the network irrespective of his or her location. The subscriber movement can be intra-region or inter-region (including inter-country). For this kind of flexibility, the subscription has to be associated with the concept of a 'public land mobile network' (PLMN). Any operational network can be said to be a PLMN. There can be one or more than one PLMN in one country. To give users the immense flexibility to be connected even when changing between PLMNs (in the same or a different country) means that communication has to take place between these PLMNs. The mobile, when entering a different PLMN, searches for its own home/serving cells. When no service is detected, then it can search in automatic mode or the desired PLMN can be selected manually.

Neighbour Cells

While in operation, a mobile is required to make evaluations regarding the quality and level of the neighbour cells. It has to decide which cells are better in terms of coverage and capacity. This is done by taking advantage of the TDMA scheme, whereby the measurements are made during the uplink transmissions and downlink reception bursts. The evaluation is done with the help of algorithms resident in the BSS. These algorithms make a decision and convey it to the mobile station.

One important thing to remember here is that every cell has its own identity code, known as the base station identity code (BSIC). Neighbour cells can have the same BSIC, so in those cases the mobile identifies the neighbour cell by a 'colour code'. Recall that SCH and FCCH play an important role in logging the mobile to the network, and these channels transmit their information on a frequency known as the *beacon* frequency. Neighbour cells may have the same beacon frequency. In such cases, the BSIC helps in distinguishing the channels of the same frequency.

2.5 RADIO NETWORK OPTIMISATION

2.5.1 Basics of radio network optimisation

Optimisation involves monitoring, verifying and improving the performance of the radio network. It starts somewhere near the last phase of radio network planning, i.e. during parameter planning. A cellular network covers a large area and provides capacity to many people, so there are lots of parameters involved that are variable and have to be continuously monitored and corrected. Apart from this, the network is always growing through increasing subscriber numbers and increases in traffic. This means that the optimisation process should be on-going, to increase the efficiency of the network leading to revenue generation from the network.

As we have seen, radio network planners first focus on three main areas: coverage, capacity and frequency planning. Then follows site selection, parameter planning, etc. In the optimisation process the same issues are addressed, with the difference that sites are already selected and antenna locations are fixed, but subscribers are as mobile as ever, with continuous growth taking place. Optimisation tasks become more and more difficult as time passes.

Once a radio network is designed and operational, its performance is monitored. The performance is compared against chosen key performance indicators (KPIs). After fine-tuning, the results (parameters) are then applied to the network to get the desired performance. Optimisation can be considered to be a separate process or as a part of the network planning process (see Figure 2.15).

The main focus of radio network optimisation is on areas such as power control, quality, handovers, subscriber traffic, and resource availability (and access) measurements.

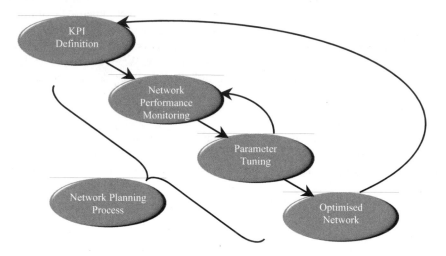

Figure 2.15 Radio network optimisation

2.5.2 Key Performance Indicators

For radio network optimisation (or for that matter any other network optimisation), it is necessary to have decided on key performance indicators. These KPIs are parameters that are to be observed closely when the network monitoring process is going on. Mainly, the term KPI is used for parameters related to voice and data channels, but network performance can be broadly characterised into coverage, capacity and quality criteria also that cover the speech and data aspects.

Key Indicators–Voice Quality

The performance of the radio network is measured in terms of KPIs related to voice quality, based on statistics generated from the radio network. Drive tests and network management systems (described later) are the best methods for generating these performance statistics.

The most important of these from the operator's perspective are the BER (bit error rate), the FER (frame error rate) and the DCR (dropped call rate).

The BER is based on measurement of the received signal bits before decoding takes place, while the FER is an indicator after the incoming signal has been decoded. Correlation between the BER and the FER is dependent on various factors such as the channel coding schemes or the frequency hopping techniques used. As speech quality variation with the FER is quite uniform, FER is generally used as the quality performance indicator for speech. The FER can be measured by using statistics obtained by performing a drive test. Drive testing can generate both the uplink and the downlink FER.

The dropped call rate, as the name suggests, is a measure of the calls dropped in the network. A dropped call can be defined as one that gets terminated on its own after being established. As the DCR gives a quick overview of network quality and revenues lost, this easily makes it one of the most important parameters in network optimisation. Both the drive test results and the NMS statistics are used to evaluate this parameter. At the frame level, the DCR is measured against the SACCH frame. If the SACCH frame is not received, then it is considered to be dropped call. There is some relation between the number of dropped calls and voice quality. If the voice quality were not a limiting factor, perhaps the dropped call rate would be very low in the network. Calls can drop in the network due to quality degradation, which may be due to many factors such as capacity limitations, interference, unfavourable propagation conditions, blocking, etc. The DCR is related to the call success rate (CSR) and the handover success rate. The CSR indicates the proportion of calls that were completed after being generated, while the handover rate indicates the quality of the mobility management/RRM in the radio network.

KPIs can be subdivided according to the areas of functioning, such as area level, cell level (including the adjacent level), and TRX level. Area-level KPIs can include SDCCH requests, the dropped SDCCH total, dropped SDCCH A_{bis} failures, outgoing MSC control handover (HO) attempts, outgoing BSC control HO attempts, intra cell HO attempts, etc. Cell-level KPIs may include SDCCH traffic BH (av.), SDCCH blocking BH (av.), dropped SDCCH total and distribution per cause, UL quality/level distribution, DL quality/level distribution etc. The TRX level includes the likes of UL and DL quality distribution.

2.5.3 Network Performance Monitoring

The whole process of network performance monitoring consists of two steps: monitoring the performance of the key parameters, and assessment of the performance of these parameters with respect to capacity and coverage.

As a first step, radio planners assimilate the information/parameters that they need to monitor. The KPIs are collected along with field measurements such as drive tests. For the field measurements, the tools used are ones that can analyse the traffic, capacity, and quality of the calls, and the network as a whole. For drive testing, a test mobile is used. This test mobile keeps on making calls in a moving vehicle that goes around in the various parts of the network. Based on the DCR, CSR, HO, etc., parameters, the quality of the network can then be analysed. Apart from drive testing, the measurements can also be generated by the network management system. And finally, when 'faulty' parameters have been identified and correct values are determined, the radio planner puts them in his network planning tool to analyse the change before these parameters are actually changed/implemented in the field.

Drive Testing

The quality of the network is ultimately determined by the satisfaction of the users of the network, the subscribers. Drive tests give the 'feel' of the designed network as it is experienced in the field. The testing process starts with selection of the 'live' region of the network where the tests need to be performed, and the drive testing path. Before starting the tests the engineer should have the appropriate kits that include mobile equipment (usually three mobiles), drive testing software (on a laptop), and a GPS (global positioning system) unit.

When the drive testing starts, two mobiles are used to generate calls with a gap of few seconds (usually 15–20 s). The third mobile is usually used for testing the coverage. It makes one continuous call, and if this call drops it will attempt another call. The purpose of this testing to collect enough samples at a reasonable speed and in a reasonable time. If there are lots of dropped calls, the problem is analysed to find a solution for it and to propose changes.

An example of a drive test plan is shown in Figure 2.16. Some typical drive tests results giving the received power levels from own cell and neighbour cells, FER, BER, MS power control, etc., are shown in Figure 2.17.

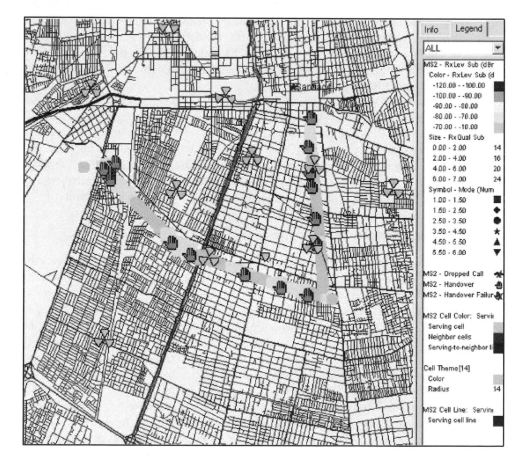

Figure 2.16 Drive test result analysis showing handovers (HO) on the path

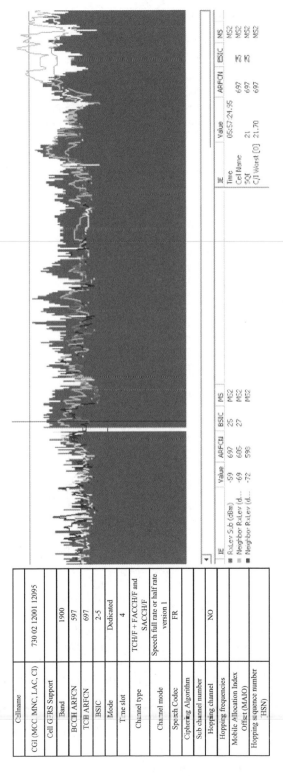

Cellname	
CGI (MCC MNC, LAC, CI)	730 02 12001 12095
Cell GPRS Support	
Band	1900
BCCH ARFCN	597
TCH ARFCN	697
BSIC	2-5
Mode	Dedicated
Time slot	4
Channel type	TCH/F + FACCH/F and SACCH/F
Channel mode	Speech full rate or half rate version 1
Speech Codec	FR
Ciphering Algorithm	
Sub channel number	
Hopping channel	NO
Hopping frequencies	
Mobile Allocation Index Offset (MAIO)	
Hopping sequence number (HSN)	

IE	Value	ARFCN	BSIC	MS
■ RxLev Sub (dBm)	-59	697	25	MS2
■ Neighbor RxLev (d...	-69	605	27	MS2
■ Neighbor RxLev (d...	-72	598		MS2

IE	Value	ARFCN	BSIC	MS
Time	05:57:24.95			
Cell Name		697	25	MS2
SQI	21	697	25	MS2
C/I Worst [0] 21.70		697		MS2

Figure 2.17 Some drive test results

Network Management System Statistics

After the launch of the network, drive tests are performed periodically. In contrast, the statistics are monitored on the NMS daily with the help of counters. The NMS usually measures the functionalities such as call setup failures, dropped calls, and handovers (successes and failures). It also gives data related to traffic and blocking in the radio network, apart from giving data related to quality issues such as frequency hopping, FER and BER, field strength, etc. An example of area-level KPI statistics is shown in Figure 2.18.

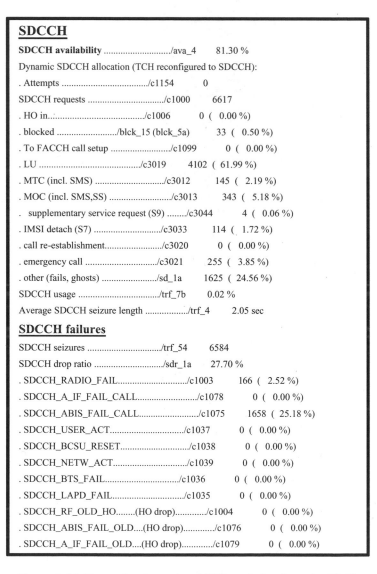

Figure 2.18 Example of area-level KPI statistics from the NMS

2.5.4 Network Performance Assessment

The performance indicators are listed below:

- amount of traffic and blocking
- resource availability and access
- handovers (same cell/adjacent cell, success and failure)
- receiver level and quality
- power control.

Coverage

Drive test results will give the penetration level of signals in different regions of the network. These results can then be compared with the plans made before the network launch. In urban areas, coverage is generally found to be less at the farthest parts of the network, in the areas behind high buildings and inside buildings. These issues become serious when important areas and buildings are not having the desired level of signal even when care has been taken during the network planning phase. This leads to an immediate scrutiny of the antenna locations, heights and tilt. The problems are usually sorted out by moving the antenna locations and altering the tilting of the antennas. If optimisation is being done after a long time, new sites can also be added.

Coverage also becomes critical in rural areas, where the capacity of the cell sites is already low. Populated areas and highways usually constitute the regions that should have the desired level of coverage. A factor that may lower the signal level could be propagation conditions, so study of link budget calculations along with the terrain profile becomes a critical part of the rural optimisation. For highway coverage, additions of new sites may be one of the solutions.

Capacity

Data collected from the network management system is usually used to assess the capacity of the network. As coverage and capacity are interrelated, data collected from drive tests is also used for capacity assessment. The two aspects of this assessment are dropped calls and congestion. Generally, capacity-related problems arise when the network optimisation is taking place after a long period of time. Radio network optimisation also includes providing new capacity to new hot-spots, or enhancing indoor coverage. Once the regional/area coverage is planned and executed in the normal planning phase, optimisation should take into consideration the provision of as much coverage as possible to the places that would expect high traffic, such as inside office buildings, inside shopping malls, tunnels, etc.

Quality

The quality of the radio network is dependent on its coverage, capacity and frequency allocation. Most of the severe problems in a radio network can be attributed to signal interference. For uplink quality, BER statistics are used, and for downlink FER statistics are used. When interference exists in the network; the source needs to be found. The entire frequency plan is checked again to determine whether the source is internal or external. The problems may be caused by flaws in the frequency plan, in the configuration plans (e.g. antenna tilts), inaccurate correction factors used in propagation models, etc.

Parameter Tuning

The ending of the assessment process sees the beginning of the complex process of fine-tuning of parameters. The main parameters that are fine-tuned are signalling parameters, radio resource parameters, handover parameters and power control parameters. The concepts that are discussed in the radio planning process and the KPI values should be achieved after the process is complete.

The major complexity of this process is the inhomogeneity of the radio network. Network planning will have used standard propagation models and correction factors based on some trial and error methods that may be valid for some parts of the network and invalid for other parts. Then, during network deployment, some more measurements are made and the parameters are fine-tuned again. Once the network goes 'live', the drive test and NMS statistics help in further fine-tuning of the parameters, and it is at this point that a set of default parameters is created for the whole network. However, as the network is inhomogeneous, these default parameters may not be sufficiently accurate in all regions, thereby bringing down the overall network quality – and leading to a reduction in revenue for the network operator.

Radio network optimization must be a continuous process that begins during the pre-launch phase and continues throughout the existence of the network.

3

Transmission Network Planning and Optimisation

3.1 BASICS OF TRANSMISSION NETWORK PLANNING

3.1.1 The Scope of Transmission Network Planning

In the GSM system, the transmission network is generally considered to be the network between the base stations and the transcoder sub-multiplexers (TSCM). The transmission network connects the radio network to the mobile switching centre (MSC), and hence, by virtue of its position and functionality, it acquires an important position in the mobile network infrastructure (see Figure 3.1).

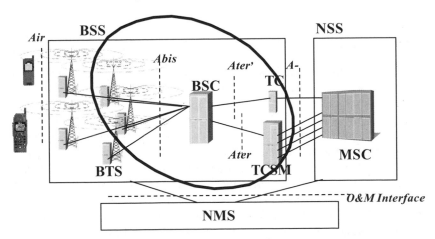

Figure 3.1 Scope of transmission network planning

Fundamentals of Cellular Network Planning & Optimisation A.R. Mishra.
© 2004 John Wiley & Sons, Ltd. ISBN: 0-470-86267-X

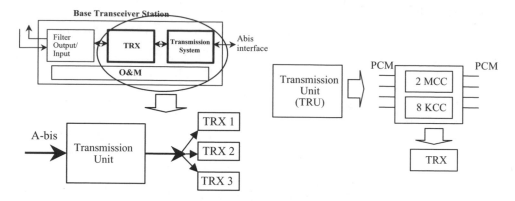

Figure 3.2 Transmission system in BTS

3.1.2 Elements in a Transmission Network

Base Station Transceiver (BTS)

An overview of the BTS is provided in Chapter 2. Here we look at the BTS from the transmission planning perspective (i.e. the transmission system within a base station) as shown in Figure 3.2.

The base station consists of a transmission unit (TRU). This unit interacts with the A_{bis} and A_{ter} interfaces. It also reallocates the traffic and the signalling channels to the correct transceiver (TRX). The TRU may be capable of making cross-connections at the 2 Mbps level and drop-and-insert functionalities at the 8 kbps level. Transmission units are capable of holding branching tables. These tables guide the transmission units on what to do with the traffic coming in from the PCM lines; e.g. should it just let it pass to the next BTS, or should it drop some traffic meant for its own TRX, etc.

Base Station Controller

The base station controller (BSC) is capable of many functions. Radio channel management is an integral function of the BSC and includes tasks such as channel management and release. It is responsible for the handover management function that is based on an assessment of signal power, signal quality at the MS in both the uplink and downlink directions, and adjustments related to the minimisation of interference in the radio network. It is also responsible for measurements related to the implementation of frequency hopping (FH) within the network. Other important functions include interaction with the base station and MSC on either side, encryption management such as storing the encryption parameters and forwarding the same to the base stations, traffic and channel management, etc. Also, the BSC–BTS signalling, which includes the LAPD, transceiver signalling (TRXSIG) and BSC–MSC signalling which is CCS7, is handled by the BSC itself. The BSC is also responsible for carrying the O&M information to the BTS.

Apart from these functions, the BSC also performs one more function of concern to transmission planners, which is traffic concentration between itself and the BTS. This means that even if there were a channel available on the air- and A_{bis} interface, the subscriber would not be able to make a call because of the blocking on the A_{ter} interface.

Transcoder and Sub-multiplexer (TCSM)

Transcoders and sub-multiplexers are used between the BSC and the MSC. Although they are present in the BSS system as shown in Figure 3.1, physically they may be situated at the MSC location.

The TCSM is capable of two functions: transcoding of speech signals and sub-multiplexing of PCM signals. The trancoder does the speech coding in the downlink direction, which is decoded in the MS, and it decodes the speech signal in the uplink direction. Sub-multiplexers are located at the same place as the transcoders and are responsible for multiplexing of the PCM links between the BSC and the MSC. As shown in Figure 3.3, three PCM signals (this figure may be different depending upon the equipment manufacturer) coming from the MSC are multiplexed to one PCM signal towards the BSC.

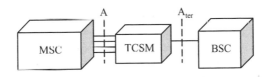

Figure 3.3 Transcoder/sub-multiplexer (TCSM)

3.2 TRANSMISSION NETWORK PLANNING PROCESS

The transmission network planning process should generate a network plan that provides a high-quality network, with an immense amount of spare capacity at very low cost. In practice, a transmission planning engineer has to use his or her expertise to balance three factors: cost, quality and capacity.

The process of transmission network planning includes five main phases before the final plan is generated, as shown in Figure 3.4. Although the process looks simple in theory, in practice it is much more complicated because there could be many iterations at each step before a final plan is ready.

The process begins with data collection, including the requirements for capacity and quality that lay the foundation of the whole process. Other essential information relates to what transmission equipment is available (including capacities), quality targets, tools for transmission planning including link budget calculations, topologies that can be used, etc.

The pre-planning phase focuses on the dimensioning aspects of the transmission network. The pre-planning starts with the inputs from the radio planning engineers. Once the number

Figure 3.4 The transmission network planning process

of base stations and their capacities are known, transmission planners will start with defining the quality targets for the network. According to the transmission products available – such as radios, base stations, base station controllers, information on the media – an imaginary network topology is defined, or the topology principles that will be used in the transmission network are formed. This gives a better idea on where the transmission network should be heading. The final network design after the whole process will be somewhat different from the one drawn up at this stage. All along the media of transmission is considered. If microwave transmission is to be used, then the link budget aspects have to considered as well, so that the link design meets the availability and quality targets.

Selection of sites is basically a radio planning issue, but the transmission planning engineers should get involved right at this early phase, so that it will be easier to find sites that have the desired lines of sight (LOS) with other sites. The line-of-sight process should be done very professionally, because once a site is *constructed* and there is no LOS with another site, that represents a big loss for the network owner.

When microwave is the chosen medium of transmission, link budget calculations are generally based on ITU recommendations. Tools used for the link budget and availability calculations should be based on the ITU/ETSI, etc., recommendations.

The detailed planning phase consists of frequency allocation/frequency planning, defining of the time slot allocation plans, routing of the PCM signals, synchronisation principles, and network management planning.

This whole process is usually an iterative one at different points, as the transmission planners try to achieve a balance between network capacity, quality and costs.

3.3 PRE-PLANNING IN TRANSMISSION NETWORK

As we saw in Chapter 2, based on the capacity, coverage and quality targets, radio network planners arrive at the number of base stations required in a given region. The task of a transmission planning engineer will be to connect these sites, which will require knowledge of the existing network infrastructure and spare capacity. The locations of the BTS, BSC and MSC sites should be known as well, along with the present and future capacity requirements for each base station.

Armed with that knowledge, the transmission planner needs to find out the required capacity of each link. This process is dependent upon the traffic generated by each base station, the ability to cross-connect and groom each site so as to make complete use of the PCM links. Also, the transmission media, future capacity and topology have their own individual impacts on the number of PCM links required to interconnect the base stations, and subsequently the base stations to the mobile.

3.3.1 One PCM Connection

One PCM line consists of 32 time slots, as shown in Figure 3.5. Time slot 0 (TS0) is used for link management, while TS1 to TS31 are used for traffic and signalling. The bit rate of each TS is 64 kbps, and depending on the type of signalling, the TCH can be 16, 32 or 64 kbps. The PCM shown in Figure 3.5 is the PCM on the A_{bis} interface. It contains the TCHs that carry the traffic from the transceiver (TRX), as well as the signalling associated with it.

		1	2	3	4	5	6	7	8
0		LINK MANAGEMENT							
1		TCH.1		TCH.2		TCH.3		TCH.4	
2		TCH.5		TCH.6		TCH.7		TCH.8	
3		TCH.1		TCH.2		TCH.3		TCH.4	
4		TCH.5		TCH.6		TCH.7		TCH.8	
5		TCH.1		TCH.2		TCH.3		TCH.4	
6		TCH.5		TCH.6		TCH.7		TCH.8	
7		TCH.1		TCH.2		TCH.3		TCH.4	
8		TCH.5		TCH.6		TCH.7		TCH.8	
9		TCH.1		TCH.2		TCH.3		TCH.4	
10		TCH.5		TCH.6		TCH.7		TCH.8	
11		TCH.1		TCH.2		TCH.3		TCH.4	
12		TCH.5		TCH.6		TCH.7		TCH.8	
13		TCH.1		TCH.2		TCH.3		TCH.4	
14		TCH.5		TCH.6		TCH.7		TCH.8	
15		TCH.1		TCH.2		TCH.3		TCH.4	
16		TCH.5		TCH.6		TCH.7		TCH.8	
17		TCH.1		TCH.2		TCH.3		TCH.4	
18		TCH.5		TCH.6		TCH.7		TCH.8	
19		TCH.1		TCH.2		TCH.3		TCH.4	
20		TCH.5		TCH.6		TCH.7		TCH.8	
21		TCH.1		TCH.2		TCH.3		TCH.4	
22		TCH.5		TCH.6		TCH.7		TCH.8	
23		TCH.1		TCH.2		TCH.3		TCH.4	
24		TCH.5		TCH.6		TCH.7		TCH.8	
25		TRXSIG1		OMUSIG1		TRXSIG2		OMUSIG2	
26		TRXSIG3		OMUSIG3		TRXSIG4		OMUSIG4	
27		TRXSIG5		OMUSIG5		TRXSIG6		OMUSIG6	
28		TRXSIG7		OMUSIG7		TRXSIG8		OMUSIG8	
29		TRXSIG9		OMUSIG9		TRXSIG10		OMUSIG10	
30		TRXSIG11		OMUSIG11		TRXSIG12		OMUSIG12	
31		x		x		x		x	

Figure 3.5 Example of a PCM

This PCM also contains the O&M signalling for the base station. Depending upon the topology, this PCM signal might also contain the bits for synchronisation and loop control.

3.3.2 PCM Requirements on the A_{bis} and A_{ter} Interface

Chapter 2 defined the blocking for the air interface. Blocking parameters need to be defined also for the A_{bis} and A_{ter} interfaces. Typically, the value of the air-interface blocking is 1–2% and that of the A_{ter} interface is 0.1–0.5%. The A_{bis} interface should have no blocking apart from no blocking between the TCSM and MSC. Radio planning engineers use the air interface for dimensioning the radio network, while transmission planning engineers use the A_{bis} and A_{ter} interfaces. The transmission capacity is determined by factors such as traffic generated at the air interface, the grooming ability of the equipments, topologies, and the locations of the BSCs and TCSMs. Spare capacity plays an important part in taking the future requirements of the network into account.

Radio planning engineers determine the number of BTS and transceivers in the network. Transmission planners start with information about the radio planning dimensioning, i.e. the number of transceivers per base station. As shown in Figure 3.5, one TRX needs 2×64 kbps time slots, while for the signalling 16, 32 or 64 kbps time slots may be required. A lower signalling rate means that more subscribers can use the network at one time. The object of dimensioning for the A_{ter} interface is to find the number of traffic channels required. Once that is done, then based on the number of channels that can be multiplexed on the A_{ter}, the capacity of the A_{ter} interface can be calculated.

Example1: Capacity requirements on the A_{ter} interface

Assume, that radio planners have decided that there are five sites of $2 + 2$ configuration under a single BSC. Air interface blocking is 2% and A_{ter}

5 sites of $2 + 2$ configuration $= 10$ cells, each having 15 TCH

Air interface blocking $= 2\%$

Using Erlang B tables, 15 TCH support $= 9.01$ Erl of traffic.

Traffic offered to the BSC $= 10 \times 9.01 = 90.1$ Erl.

If A_{ter} blocking probability is 0.1%, then the number of traffic channels supported $= 117$ (approx.)

If the number of traffic channels that can be multiplexed on the $A_{ter} = 120$

Then A_{ter} interface capacity would be $= 117/120=0.975 \approx \mathbf{1}$ El

3.3.3 Equipment Location

The base station and TCSM locations need to be decided during the nominal planning phase itself. As shown in Figure 3.3, the transmission capacity required between the TCSM and MSC is more compared with the connection towards the BSC. Thus, if the TCSMs were placed physically near the MSC, it would save transmission costs. When it comes to BSC locations, they can be located far from the MSC or co-located with the MSC. Both the remote and co-located BSC locations have their own advantages and disadvantages. A remote BSC location is chosen to cater for the needs of the huge number of BTSs that are located very far from the MSC. In this way, transmission costs are saved, though there may be heavy usage of the A_{bis} links. This is usually preferred for rural regions. A co-located BSC is generally chosen for metropolitan areas where there is a heavy concentration of traffic.

3.3.4 Network Topology

Selection of the most appropriate network topology can lead to huge savings in the transmission network. There are four main topologies, as shown in Figure 3.6.

- Point-to-point topology: As the name suggests, this is a point-to-point connection between the base stations and the BSC. This is quite prevalent during the early phase of network evolution as the implementation is easy and quick. However, the capacity may or may not be fully utilised. Also, equipment protection measures are an additional expense.

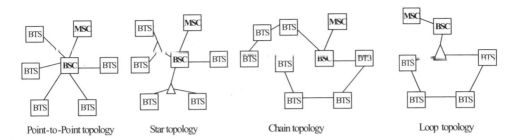

Figure 3.6 Transmission network topologies

- Star topology: This topology resembles a star – hence the name. With this arrangement the capacity can be utilised in a better way by using cross-connecting equipment. This topology is also simple to implement, but if a node fails for any reason then the traffic coming through it towards the BSC is lost. Again, protection is expensive in this kind of topology.

- Chain topology: This is generally used to connect the sites that extend from one region/city to another region/city. A typical example is connection of sites on a highway. The capacity of the links is usually fully utilised. However, the last link of the chain can be vulnerable.

- Loop topology: This is an extension of the chain topology, with the last link of the chain being connected back to the BSC. Although this requires an additional link in the chain, it provides protection without additional hardware redundancy. Also, there is a requirement for an additional node, which is able to control loop protection.

In practice, a transmission network will employ all of the above technologies. Transmission planning engineers have to balance the technological and cost aspects, and the strategic importance of the site before a decision is made.

3.3.5 Site Selection and Line-of-Sight Survey

Site Selection

The site selection process is described in Chapter 2 because it is predominantly a radio network planning engineer's domain, but with the active involvement of transmission planning engineers. The types of site that are preferred by radio and by transmission planning engineers have opposite characteristics. Radio planners prefer sites that are not too high, so as to prevent interference problems from other sites, while transmission planning engineers prefer sites that are very high, so that connectivity to a large number of other sites is possible. However, both agree that a site needs to have connectivity, so there is likely to be some compromise on the height of the site to be selected in a given region. Sometimes, the important sites chosen by the radio planners are low or are between high building structures, and in such cases repeaters are used by transmission planners. These repeater sites are usually on high buildings having only transmission equipment.

Line-of Sight Survey

Connectivity is a very important criterion for site acceptance for both transmission planning engineers and radio planning engineers. Transmission planners make sure that one site is visible from the next site. It is not sufficient for one site merely to be visible from another but LOS criteria must be clear before a site is accepted. This process may also lead to construction of the path profile, especially with longer hops. This clearance criterion is based on the concept of a *Fresnel zone*. A typical Fresnel zone and path profile are shown in Figure 3.7.

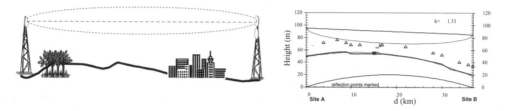

Figure 3.7 Fresnel zone and path profile analysis

3.3.6 Radius of the Fresnel Zone

Typically, the Fresnel zone can be defined as an area that is covered by an imaginary ellipsoid (a three-dimensional ellipse, as shown in Figure 3.8) drawn between the transmitting and receiving antennas in such as way that the distance covered by the ray being reflected from the surface of the ellipsoid and reaching the receiving antenna is half a wavelength longer than the distance covered by the direct ray travelling from the transmitting to the receiving antennas, i.e. $d_1 + d_2 = d + \lambda/2$.

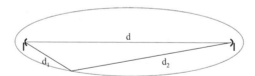

Figure 3.8 Radius of the Fresnel zone

The Fresnel zone is dependent upon two factors: frequency of transmission and the distance covered. Mathematically, the radius of the Fresnel zone can be calculated as:

$$F_1 = 12.75(d_1^* d_2/f^* D)^{1/2} \tag{3.1}$$

where F_1 is the radius of the first Fresnel zone, f is the frequency of the transmitted signal in GHz, and D, d_1 and d_2 are distances in km (as shown in Figure 3.8).

In most cases, the radius of the first Fresnel zone should be kept 100% clear. This means that there will be no reduction in signal level due to the terrain. If this first Fresnel zone were blocked, then the percentage of losses would be proportional to the percentage blocking in the Fresnel zone. In practice, transmission planning engineers try to keep some 'clearance' between the Fresnel zone and the highest obstacle in the radio path.

3.3.7 Microwave Link Planning

This constitutes a very important part of the transmission planning process in networks that use microwave transmission. The aim of microwave link planning is to achieve the desired performance at lowest cost. The two major aspects of link planning are the link budget and propagation phenomena.

Link Budget Calculations

The aim of link budget calculations is to find out what path length would be suitable for getting the desired signal level. This received signal level is then used for calculation of the fade margin. Fade margin can be defined as the margin required for countering the effects of a decrease in the signal level (also known as 'fading') due to atmospheric conditions, etc.

The first factor required in the calculation of the fade margin is the free-space loss, which is the same as the one introduced in Chapter 2. The phenomenon of free-space loss (FSL) stands true for microwave planning, except that now the distance changes from a few hundred metres to a few kilometres. FSL is dependent on the frequency of the microwave signal and the distance covered:

$$\text{FSL } (L_{\text{fs}}) = 92.5 + 20 \log d + 20 \log f \tag{3.2}$$

The second factor required is the gain of the antenna. The antenna should have sufficient gain so as to receive the signal at a desired level. The gain of a microwave antenna is given as:

$$G = 10 \log(4\pi{}^{*}A^{*}e/\lambda^{2}) \tag{3.3}$$

where G is the gain of the antenna in dB, A is the area of the antenna aperture, e is the efficiency of the antenna, and λ is the wavelength (same units as A). Equation 3.3 can also be written as (for parabolic antennas):

$$G = 20 \log D_{\text{a}} + 20 \log f + 17.5 \tag{3.4}$$

where G is the gain of the antenna in dB, D_{a} is the diameter of the antenna in metres, and f is the frequency of operation in GHz.

Based on the above, the hop loss L_{h} can be calculated as:

$$L_{\text{h}} = L_{\text{fs}} - G_{\text{t}} - G_{\text{r}} + L_{\text{ext}} + L_{\text{atm}} \tag{3.5}$$

where G_{t} is the gain of the transmitting antenna, G_{r} is the gain of the receiving antenna, L_{ext} is extra attenuation (due to radome, etc.), and L_{atm} stands for atmospheric losses due to water vapour and oxygen. The received signal level P_{rx} can then be calculated as:

$$P_{\text{rx}} = P_{\text{t}} - L_{\text{h}}. \tag{3.6}$$

Mathematically, the fade margin can be described as the difference between the received signal power and the receiver threshold (R_{xth}):

$$FM = P_{\text{rx}} - R_{\text{xth}}. \tag{3.7}$$

Example 2: Calculation of fade margin

Calculate the fade margin of a microwave link whose dimensions are as follows:

Hop length = 10 km
Frequency = 15 GHz
Antenna diameter = 0.6 m
Transmit power = 20 dBm
Extra attenuation = 0 dB
Atmospheric attenuation = 0 dB
Receiver threshold = −75 dB

Using equations 3.2 to 3.7:

FSL (L_{fs}) = 92.5 + 20 log (10) + 20 log (15) = 136.02 dB

Gain of antenna (G_t and G_r) = 20 log (0.6) + 20 log (15) + 17.5 = 36.58 dBm

Hop loss (L_h) = 136.02 − 36.58 − 36.58 − 0 − 0 = 62.86 dB

Received signal level (P_{rx}) = 20 − 62.86 = − 42.86 dB

Thus, fade margin (FM) = −42.86 − (−75) = 32.14 dB

Propagation Phenomena

In this section, we try to look into the factors that would affect the received signal level at the receiving antenna. As seen above, there are losses in the received power between the transmitting and the receiving antennas and this loss has been termed 'free-space loss'. Apart from this, there are many other factors such as weather conditions, rain, terrain profile, etc., that need to be considered before the link design can reach the implementation stage. First, however, the principles of propagation should be understood.

When a signal passes from one medium to another, the trajectory and velocity of the signal change, owing to the refraction that takes place in the atmosphere. The ratio of the velocity of the signal in free space to its velocity in another medium is known as the refractive index (η) of that medium. This term signifies the *refractivity* of the medium:

$$\eta = V_{fs}/V_m \tag{3.8}$$

where V_{fs} is the velocity of the signal in free space and V_m is its velocity in the other medium through which it is passing.

The Earth's atmosphere is a mixture of gases. The refractive index of the atmosphere is dependent upon three factors: pressure, humidity and water vapour. As we move up higher into the atmosphere, all these three change. On a mountain you can feel the change in pressure and humidity and, of course, it becomes cooler. Thus, the refractive index of the atmosphere, which is normally considered to be close to unity, has to be calculated taking these three effects into account. The value is approximately 1.00045.

Another term used is *radio refractivity*, N, which is defined as:

$$N = (\eta - 1)*10^6 \tag{3.9}$$

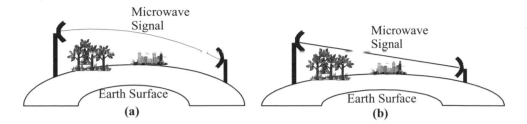

Figure 3.9 Microwave signal trajectory

Upto 300 GHz, this equation is also written in terms of pressure, humidity and temperature as:

$$N - 77.6(P/T) + 3.73^*10^{6*}(e/T^2) \tag{3.10}$$

where e is the partial pressure of water vapour (millibars), T is the absolute temperature in kelvin, and P is the pressure (millibars).

Owing to changes in the pressure, temperature and humidity at every single point in the atmosphere, there will be changes in the radio refractivity at each point. We can assume that a small change in height, dh, will bring about a small change in radio refractivity, dN. Thus, the path of the signal is not a straight line but is in the shape of an arc. Hence, in reality, the trajectory of the signal in relation to the surface of the earth is as shown in Figure 3.9(a). This kind of situation (i.e. two arcs, one for the microwave signal and one the Earth's surface) is complicated to analyse mathematically. To make the analysis easy, one of the two arcs is made into a straight line as shown in Figure 3.9(b), where the microwave signal path is made a straight line, so changing the Earth's radius in the process.

Transmission planning engineers have to consider another parameter, known as the k-factor. This factor is the one used to calculate the antenna heights and is defined as the ratio of the *effective* radius of the Earth, ka, to the actual radius of the Earth, a:

Thus
$$k = ka/a. \tag{3.11}$$

This k-factor can also be derived using the above analogy and using equation 3.10, because this factor is dependent upon the curvature of the microwave signal which in turn is dependent upon the pressure, humidity and temperature at that point on the Earth's surface. As these factors all vary with height, it can as well be said that the k-factor also varies with height:

$$k = 157/[157 + (dN/dh)]. \tag{3.12}$$

During standard conditions of temperature and pressure, the value for dN/dh is considered to be -40 N/km. Thus, the k-factor turns out to be 4/3. This is the reason why transmission planning engineers plan microwave links using the k-value of 1.33. From the k-value, it is easy to predict the trajectory of the microwave signal if the path profile is known. Using the concept of the path profile and the trajectory of the microwave signal, along with the concept of the Fresnel zone, the minimum distance between the highest obstacle and Fresnel zone can be determined. This then leads to the calculation of antenna heights on two sites.

Pressure, humidity and temperature all change in the 24-hour cycle of the day. This leads to a continuous variation of radio refractivity, N, and thus the k-factor during the day.

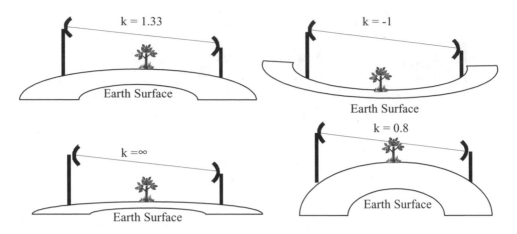

Figure 3.10 Variation of the *k*-factor

Transmission planning engineers should make themselves familiar with the known climatic changes in the city/region under consideration. In practice, the Earth's curvature remains constant so only the trajectory varies with the value of *k*, but it is easy to understand and analyse the phenomenon if the trajectory is kept constant (i.e. a straight line) and variation is made to the surface of the Earth. Variation in the values of *k* and subsequent variation in the Earth's curvature is shown in Figure 3.10.

The condition $k = 1.33$ is for the standard atmosphere variability (dN/dh) of -40 N/km. Under this condition, the Fresnel zone should be 100% clear. When the variability increases beyond -40 N/km, then the path will experience sub-refractive conditions. Owing to this increase and the subsequent variation in the *k*-factor, the Earth's surface would look to be flat, thereby taking the signal closer to the surface of the Earth. This would mean that the obstacles that were 'clear' before the occurrence of this condition now start to obstruct the Fresnel zone. This condition is shown as $k = 0.8$. When the variability reduces substantially from the -40 N/km value, then the path experiences super-refractive conditions which means that the Earth's curvature 'increases' substantially. If the value of dN/dh can be reduced to -157 N/km, then the k-factor tends to infinity. This would mean that the Earth would look to be flat, and so the microwave signal also. Further reduction would mean a negative value of *k*, thereby increasing the probability of ducting taking place in the microwave link.

Multipath Propagation

When the microwave signal leaves the transmitting antenna, it can take different paths to reach the receiving antenna (see Figure 3.11). One of the paths taken is a straight line or the Fresnel zone. Other paths may include reflections from the atmosphere or the surface of the Earth. This is called multipath propagation and results in more than one signal reaching the receiving antenna. This may result in reduction of the signal level if the phase difference is an odd multiple of half a wavelength. The severity of the reduction of the signal value will depend upon the reflected signal level. Hence, the signal level arriving at the receiving antenna is the sum of all the signals arriving at that antenna.

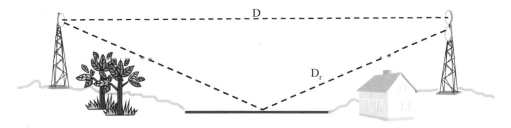

Figure 3.11 Multipath propagation

Generally, most surfaces reflect the signal, but the losses incurred during the reflections can make the reflected signal so weak that it is either not able to reach the receiving antenna or is not able to make any impact on the received signal level. The term *multipath fading* is used for the fading of the signal caused by (Earth) surface reflections, while fading caused by reflections/refractions from the atmosphere is called *selective fading*.

The layers in the atmosphere that reflect the microwave signal will cause fades in the received signal that can range from tens of decibels to 100 dB/s, but the duration of these fades is only a fraction of a second. This is also known as the layering phenomenon. The layering of the atmosphere can be caused by things like radiation nights, advection, subsidence, frontal systems, etc. *Radiation nights* occur over land when a cold night follows a warm day, resulting in the surface of the Earth cooling and the air at the surface cooling down faster than the air higher up in the atmosphere, causing a temperature inversion. *Advection* is caused by a high-pressure system moving land-based warm dry air over the cooler moist air above the sea, causing a temperature inversion. *Subsidence* is the descent of dry air in a high-pressure system that becomes heated by compression and spreads out over the cooler moist air, causing a temperature inversion. Localised inversion can also be caused by a *frontal system* driving cold air beneath warm air. Thus, the layering phenomenon is usually caused in hot climatic regions and on land near large water bodies.

Multipath fading has a substantial impact on signal strength at frequencies up to 10 GHz. With increasing frequency the impact of this kind of fading diminishes and signal fading is more to do with rain effects.

The effects of multipath fading can be countered by using *diversity*. Usually space diversity is used, i.e. two receiving antennas separated from each other in space. When, D_{r1} minus D is an odd multiple of the wavelength, multipath fading will occur. Thus, the difference between the two diversity (at the receiving site) antennas should be such that $D_{r1} - D_r$ is an odd multiple of half wavelength (Figure 3.12).

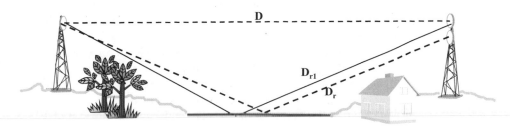

Figure 3.12 Multipath fading

Fading of the signal level can be broadly classified into two kinds: flat fading and frequency-selective fading. These and other forms of fading are now discussed.

Flat Fading

Flat fading is 'flat' across the transmitted spectrum, which means that, in the absence of any other source that would reduce the signal level, the fading is a function of thermal noise only. There are two forms of flat fading: ducting, and fading due to rain attenuation.

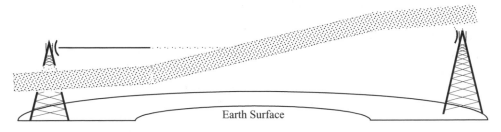

Figure 3.13 Ducting

Ducting Effect

In certain conditions the atmosphere acts as a duct, and thereby becomes capable of trapping the signal within itself. The signal departs from its original direction and moves in the direction of the duct, as shown in Figure 3.13.

Once the signal is trapped in the layer, it propagates as if travelling inside an optical fibre cable and going in the direction of the waveguide. The signal gets trapped only when the critical angle is exceeded, so the signal should enter the duct at an angle less than the critical angle. The ducting phenomenon also depends upon the intensity of the duct, dimensions, and the frequency of operation. Ducts do not need to travel upwards as in the figure, but may also travel downwards towards the surface of the Earth.

Rain Attenuation

Rainfall attenuates a microwave signal. The level of attenuation is dependent upon factors such as the frequency of operation, the polarisation of the signal, the path length, and the rainfall rate. Other rain factors that influence attenuation include drop size, shape and distribution, terminal velocity and temperature.

Rain starts having an impact at frequencies above 10 GHz, and the attenuation becomes critical beyond 15 GHz. The higher the frequency of operation, the more significant the attenuation due to rain. For links working at 12–15 GHz the attenuation may reach 10–12 dB/km, which is significantly high.

Transmission planning or microwave link planning engineers should remember that the rainfall rate varies with time and location. Thus, the accuracy of the calculations depend upon how the rainfall was recorded and how stable the cumulative distribution remains over a given period. In general, it has been seen that one year of data has 0.1% of error, while two years of data contains 0.01% of error; if the data are collected for a period of 10–15 years, the error rate reduces to 0.001%. In the absence of any data, it is recommended that the ITU maps be used instead.

The attenuation (A) due to rain is given by:

$$A = \alpha \cdot R^{\beta} \tag{3.13}$$

where R is the rainfall rate and α and β are constants. Both the parameters α and β are defined for spherical drops and they are independent of polarization. The value of these constants depends upon the drop size distribution, canting angle, terminal velocity and the properties of water. During network design these values are not easily available, so transmission planning engineers are recommended to use the ITU recommendation 838 as it gives the values for these parameters for both horizontal and vertical polarisations.

Attenuation also increases with distance and frequency (see Figure 3.14). Thus, in summary, attenuation increases with distance, frequency and rain rate.

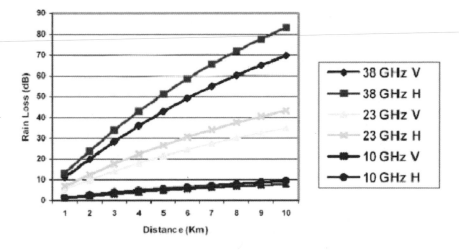

Figure 3.14 Attenuation due to rain

How does rain cause attenuation? Energy is lost from the microwave signal when the raindrops scatter or absorb it. Scattering occurs when a wave impinges on an obstacle without being absorbed, so the effect of the scattering is to change the direction of travel. If the particle is small compared with the incident wavelength, the process is described by the Rayleigh scattering phenomenon ($<10\,\text{GHz}$); when the particle is of similar size to the incident wavelength, the process is described using the Mie scattering theory ($>10\,\text{GHz}$).

Because of rainfall, wet radomes may cause additional losses that need to be accounted for in the link budget calculations. These losses can be caused by reflection, absorption and/or scattering. If the antenna (and its radome) is quite old, the losses may be even greater than a couple of decibels.Thus, attenuation due to rain can be summarized as follows:

- As the distance increases, the attenuation increases for a given rainfall rate.

- With a constant distance, the attenuation the microwave link increases with the rainfall rate.

- Vertically polarized links experience less degradation in signal level than do horizontally polarised links.

- Extra attenuation losses (e.g. wet radome losses) need to be taken into account.

Frequency-selective Fading

With ducting, the entire signal is trapped and (usually) lost. In contrast, with frequency-selective fading only a part of the signal may be lost. The atmosphere acts in a way that the beam is not trapped completely, but only deflected. The signal may still reach the receiving antenna, but rarely in phase with the direct-path signal.

The total fading (P_{TF}) in a system can be considered to be a sum of flat and selective fading:

$$P_{TF} = P_{FF} + P_{SF} \qquad (3.14)$$

where P_{TF} is the probability of fading (combined effect of flat and selective fading), P_{FF} is the probability of fade due to flat fading, and P_{SF} is the probability of fade due to selective fading.

Other Types of Fading

Apart from flat fading and frequency-selective fading, there are some other types of fading that transmission/microwave planning engineers should be aware of.

k-fading
As explained and shown above (see Figure 3.10), there is fading of the microwave signal due to variations in atmospheric conditions – pressure, temperature and humidity – which

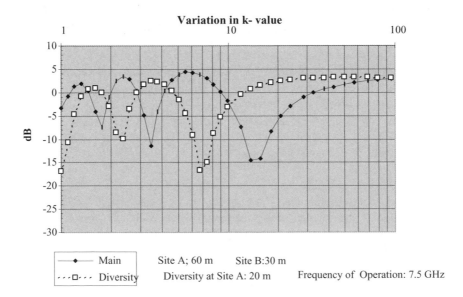

Figure 3.15 Received field (dB) at a station as a function of k

results in variation of the k-factor. This is called k-fading. An object which could be considered as clear from the first Fresnel zone may well start acting as an obstacle with changing atmospheric conditions i.e. with changing k-factor. For this reason, transmission/microwave planning engineers should plan the link for clearance not only for standard conditions (i.e. $k = 1.33$), but also for non-standard conditions such as $k = 0.6$, etc. The variation of the received signal power (with diversity antenna) versus the k-value is shown in Figure 3.15.

Diffraction Fading

Diffraction occurs when a microwave signal grazes an object and in consequence loses (disperses) energy. This is known as diffraction fading. The amount of energy lost is dependent on the type of surface or object. Objects may act as obstacles (i.e. interfere in the first Fresnel zone under standard atmospheric conditions when $k = 1.33$), or may act as obstacles under changing atmospheric conditions (i.e. during k-fading). Fading due to diffraction is primarily of two types smooth-sphere or knife edge (see Figure 3.16):

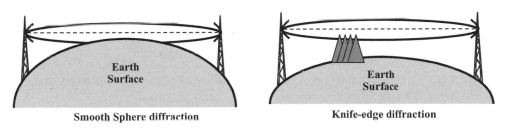

Figure 3.16 Types of diffraction

- Smooth-sphere diffraction: This takes place when the microwave signal passes over the surface of the Earth (e.g. over water). Microwave links having path lengths into tens of kilometres, passing over vast plain land or a water body, may display this when there are changes in the atmospheric conditions (i.e. k-fading). The losses due to smooth-sphere diffraction may be as high as 12–15 dB.

- Knife-edge diffraction: The losses that occur due to grazing of the microwave signal over bushes, trees or forests are called knife-edge diffraction. This may happen during normal functioning of the links and be due to variation in atmospheric conditions (k-fading). There are a few methods calculating the losses due to this kind of diffraction, the major ones being the Epstein–Peterson method, the Bullington method and the Deygout method. Each of these methods has its own merits, so transmission planners are advised to see their benefits in particular conditions before applying them. In general, it is assumed that diffraction loss due to knife-edge obstacles can be approximately 6 dB.

Atmospheric Fading

Fading due to the atmosphere may include fading due to gases (e.g. oxygen) and water vapour. Losses in the signal level due to oxygen and water vapour are predominant beyond a frequency of 15 GHz. The attenuation due to oxygen peaks at about 60 GHz, and the losses can be as high as 12–15 dB/km. For water vapour, the peak is below 45 GHz, and attenuation can be as high as 0.5–2.0 dB/km.

Other Factors Affecting Fading

Transmission/microwave planning engineers should also take into account the effect of snow, sand, dust, etc. Snow affects the performance of the link in two ways. The water content of the snow degrades the signal level, and snow on the radome degrades performance of the antenna. Losses due to snow can be as high as three times the equivalent amount due to rainfall, depending on the water content of the snow. Dry snow causes relatively less attenuation. Hail, which is a mixture of air, water and ice, can cause a loss up to 6 dB/km. The effect of sand and dust is relatively less, with the signal degrading by 1.37 dB/km at 37 GHz.

3.3.8   Design Principles for a Microwave Link

Climatic conditions cannot be predicted accurately for all times, but if the conditions are observed over a few years then microwave links can be designed with fair accuracy. The following are some design rules/principles that will act as guidance:

- The fade margin should be sufficiently large. It should be able to handle degradation in the signal level due to rain, multipath fading, k-fading, etc.

- The clearance should be checked for various fade margins – $k = 1.33, 0.6, 0.5$, etc.

- The first Fresnel zone should be free of obstacles for $k = 1.33$.

- For hops respectively under or over 15 km, the clearance should be zero for k-factors of 0.3 and 0.5.

- Over-water hops should be avoided. If they are unavoidable, the antenna heights should be chosen such that the reflection point is not falling over water. Another method to avoid the reflection point falling over water is to place the antenna such that the reflected ray is blocked (e.g. place the antenna on the far side of the terrace so that the reflected ray is blocked by the roof of the building itself).

- Choose higher antennas if the probability of k-fading is high in the region.

- In regions where the ducting phenomenon is high, choose higher antennas, as it is easier to move an antenna down rather than up on the tower (especially if the antenna is placed near the top of the tower).

3.3.9   Error Performance and Availability

Once the link has been designed, its performance is measured against error performance and availability targets. The error performance objective defines the parameters to be measured against certain standard values, while the link availability is characterised on the basis of the error performance parameters. These objectives are defined in ITU recommendations ITU-R G.826.

Example 3: Error performance calculations for radio relay system

Calculated hop from		Site A	to	Site B
Radio frequency	8.1 GHz vertical			
Hop length	35.0 km			
Latitude	6 N			
Longitude	80 E			
Percentage pL	50.0 %			
Geoclimatic factor	1.41E-003			
Rainrate (0.01%)	120.0 mm/h			
Station heights (ref. level)		200.0 m		200.0 m
Antenna heights (above ref. level)		50.0 m		50.0 m
Feeder lengths		60.0 m		60.0 m
Feeder loss/100 m		6.1 dB		6.1 dB
Feeder type		EW 77		EW 77
Antenna diameters		3.7 m		3.7 m
Transmitter output power (P_{tx})		27.0 dBm		
Free space loss (L_o)		141.6 dB		
Additional terrain loss (L_{ad})		0.0 dB		
Antenna branching loss(L_{br})		5.0 dB (SD,HSB)		
Feeder losses (L_{c1})		4.2 dB		
Connector Loss (L_{c2})		4.2 dB		
Antenna gains(G_{a1} and G_{a2})		47.3 dBi		
Hop loss in non-faded state (L_{ho})		60.3 dB		
Received unfaded power (P_{rx})		−33.3 dBm		
Receiver threshold power (P_{rxth})		−76.5 dBm		
(at BER 10 −3)				
Flat fading margin (M)		43.2 dB		
Calculated flat outage time (p_{fm})		0.0283 %		
Total outage time (non-diversity) p		0.0283 %		
Annual unavailability due to rain		0.0003 %		
With space diversity :				
Diversity antenna diameter (D_d)		3.7 m		
Antenna gain (G_d)		47.3 dBi		
Antenna spacing (s)		10.0 m		
Diversity improvement (I_{fm})		34.9		

3.4 DETAILED TRANSMISSION NETWORK PLANNING

3.4.1 Frequency Planning

A microwave link is rarely situated in an isolated environment. There are usually many microwave links in a given region, of the same and/or different network operators. Then there are links that are not related to cellular networks radiating frequencies that may cause problems in these networks, e.g. radar systems. Owing to the presence of so many different sources of radiation in a region, it is very important to construct correct frequency plans for one's own network. The process usually starts with application for a frequency being made to the local governing bodies. Owing to congestion, the number of frequency bands given to a cellular operator are usually less than desired, making the need for an efficient (i.e. interference-free) frequency plan even more necessary.

Before we go into frequency planning, let us see the various kinds of interferences and sources of interference that may exist in a network:

- intra-system interference (noise, imperfections, etc.)
- inter-channel interference (adjacent/co-channel, etc.)
- inter-hop interference (front-to-back, over-reach)
- external interference (other systems, radar, etc.).

For frequency planning in the microwave network, inter-channel interference and inter-hop interference are of utmost important for transmission planners. Intra-system interference is caused by noise generated within the system and can be removed by good system design. Inter-channel interference consists of interferences caused by co-channel and adjacent channels. Co-channel interference means that some signals from identical frequency bands are reaching the microwave antenna. However, co-channel interference can only be cross-polar, i.e. the signals will be 90 degrees out (e.g. one is horizontally polarised while other is vertically polarised). Adjacent-channel polarisation means that both the frequency bands of both the signals are next to each other (e.g. one can be 18.1–18.4 while the other is 18.2–18.6 GHz). Unlike co-channel interference, adjacent-channel interference can be both co-polar and cross-polar. These are shown by signals 1, 3 and 4 in Figure 3.17. Using non-consecutive frequency bands in the same or adjacent links can prevent this kind of interference.

Inter-hop interference is caused by an undesired signal arriving at the receiving antenna. Using different frequency spots at potential sources of interference can prevent this kind of

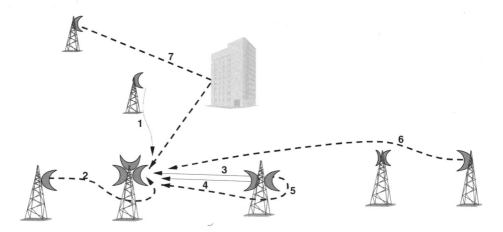

Figure 3.17 Sources and types of interference: (1) co-channel/adjacent channel from different hops; (2) opposite hop front-to-back reception; (3) adjacent channel, same hop; (4) cross-polarisation, same hop; (5) front-to-back radiation; (6) over-reach; (7) terrain reflection

interference. Techniques such as using different polarisation in subsequent links, and tilting of antennas, are remedies.

Sometimes interferences are detected in a network, even though frequency plans are perfect. This may be due to possible reflections that might be taking place in the network. In these kinds of situations, again, techniques like antenna tilting, or reducing power to the transmitting antenna (interfering source), can be used.

Example 4: Interference Calculation (Over-reach): This example calculates the signal to interference ratio (S/I) and threshold degradation due to over-reach interference.

Two links are assumed to be operating at the same frequency. The links, however, may belong to the same or a different network, as shown in Figure 3.18.

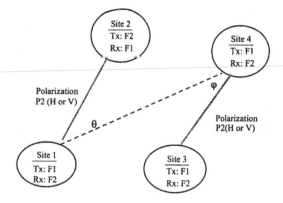

Figure 3.18 Interference calculation

The following aspects are considered:

- *Site 4 will receive interference signals from site 1 and vice versa. The interference may cause degradation of the receiver threshold to an extent dependent on the received S/I ratio at the site.*

- *Since the antennas at sites 1 and 4 are oriented towards sites 2 and 3 respectively, an antenna discrimination will result, depending on the off-axis angles θ and φ.*

- *Also contributing to this discrimination could be a difference in polarization of the two antennas at sites 1 and 4.*

- *Furthermore, since the antennas at sites 1 and 4 are mounted at heights dependent on the paths between sites 1–2 and sites 3–4 respectively, the path between sites 1 and 4 may cause additional attenuation over and above the free-space loss, because of obstructions. As a first analysis, however, the additional loss may be assumed to be zero.*

(Note: Along with the above form of interference, other sources may also contribute to interference, as shown in Figure 3.17.)

The following factors should be taken into account:

- *antenna discrimination*

- *cross polarisation discrimination*

- *diffraction loss due to obstructions in the interfered path.*

Table 3.1

System Details	SITE 1	SITE 4	Units
Transmitter Output	Ptx1	Ptx4	dbm
Feeder Loss	Lf1	Lf4	db
Connector Loss	Lc1	Lc4	db
Received Carrier Power	Prx1	Prx4	dbm
Polarization	P1	P4	H, V
Threshold Degradation	Y1	Y4	db
Receiver Threshold	Th1	Th4	dbm
Antenna Gain	G1	G4	dBi

Geographic Details	SITE 1	SITE 4	Units
Distance	D	–	Km
Off Axis Angle	θ	ϕ	deg

Polarization	SITE 1	SITE 4
HH	HH1	HH4
VV	VV1	VV4
VH	VH1	VH4
HV	HV1	HV4

The inputs required are tabulated in Table 3.1.
The interference level at site 4 (I_4) is given by:

I_4 = Received signal level at site 4, receiving from site 1 – combined antenna
discrimination at both ends

= TX power level of 1 + antenna gain at 1 – wave-guide loss at 1 – Attenuation
over path 1 to 4 (FSL + additional Terrain Loss) + antenna system gain at 4 –
wave-guide loss at 4 – combined antenna discrimination at both ends.

or

$$I_4 = P_{\text{tx}1} + G_1 - (L_{\text{f}1} + L_{\text{c}1}) - \text{FSL} - \text{additional terrain loss}$$
$$+ G_4 - (L_{\text{f}4} + L_{\text{c}4}) - \text{combined antenna discrimination.}$$

The combined antenna discrimination can be derived from radiation pattern details of site
1 and site 4 antennas, depending on the polarisation combination (HH, HV, VH or VV)
between sites 1 and 4.

Example 5: Frequency Plan

Taking a cue from Figure 3.17, an example of a possible frequency plan is shown in
Figure 3.19. Interferences 1, 3 and 4 (from Figure 3.17) can be cured by using different
frequency bands and polarisations. Interferences from 2, 5 and 6 can be removed by using
a polarisation-altering technique (i.e. altering the horizontal and vertical polarisations).

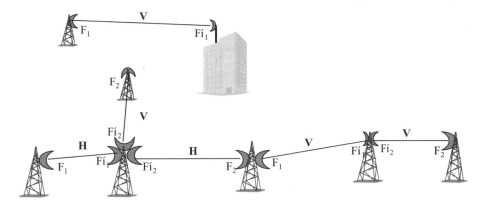

Figure 3.19 Interference calculation

Interference due to signal 7, which is detected only (usually) after the network is live, can be removed by using either different frequency bands or polarisation. As mentioned previously, other techniques such as antenna tilting can also be used; but as this requires the link to be non-functional, other techniques such as power reduction can be used.

It is always preferable to have more frequency bands at one's disposal for better frequency planning, but congestion of the frequency bands, especially at lower frequencies, usually makes this difficult.

3.4.2 Time-slot Allocation Planning

Understanding of time-slot allocation principles and methodology is crucial for the efficient utilisation of the PCM lines available. These principles are also crucial for the traffic distribution at the base station.

First let us examine the movement of one PCM or 2 Mbps from the MSC to the MS. One PCM was shown in Figure 3.5. On the A-interface, one 2 Mbps consists of 32 time slots. TS0 is used for link management. Each of these time slots is a 64 kbps channel. The next stop for the 2 Mbps signal is the transcoder/sub-multiplexers, which convert these 64 kbps signals into 16 kbps traffic channels. The common channel signalling passes through unchanged. Sub-multiplexer then maps them into a single 2 Mbps channel and sends them to the base station. The traffic at the base station determines the number of the TRXs there. Traffic channels remain at 16 kbps on the A_{bis} interface along with the addition of some signalling channels required for the transceivers and the base station signalling. Beyond the base station (i.e. on the air interface), the 16 kbps channels are converted into the 13 kbps traffic and 3 kbps inband signalling channels.

As seen before, each TRX needs two time slots of 64 kbps and one signalling channel of 16 kbps for each TRX. One signalling channel is required for each base station, known as the base station control or BCF signalling. One TRX has a handling capacity of 12 transceivers (this figure may vary from equipment to equipment).

Basically, TS allocation planning involves the allocation of the traffic channels (and time slots) for traffic and signalling apart from the pilot bits (for synchronisation) and loop control bits. There are a number of ways in which the transmission planner can allocate the time-slot and signalling channels. The allocation of the time slots should be done in such a way that there is a possibility for an upgrade from, for example, a $1 + 1 + 1$ configuration to $3 + 3 + 3$. While doing such an upgrade, there should be minimal changes made in the allocation table and the branching tables. Also, there should always be space for the management bits for the loop protection, synchronisation, etc.

Time slots can be grouped either in linear fashion or in block fashion. Linear allocation means that the next free time slot is allocated for the new TRX. The main advantage of this technique is that it is simple to plan. But as the network starts varying, and a few changes are being made, it becomes difficult to keep a track, which is not very suitable for big and complicated networks.

Block allocation of the time slots in the whole PCM is divided into standard blocks and each of these blocks is then assigned to the respective TRX. This scheme of allocation is easy to upgrade. As the spare capacity is distributed along the whole 2 Mbps signal, it is difficult to add a whole new base station.

One way of doing time-slot allocation was shown in Figure 3.5, where all the traffic channels are allocated time slots 1–24 while all the signalling is confined to the lower part of the 2 Mbps signal. Another way of doing time-slot allocation planning is shown in Figure 3.20,

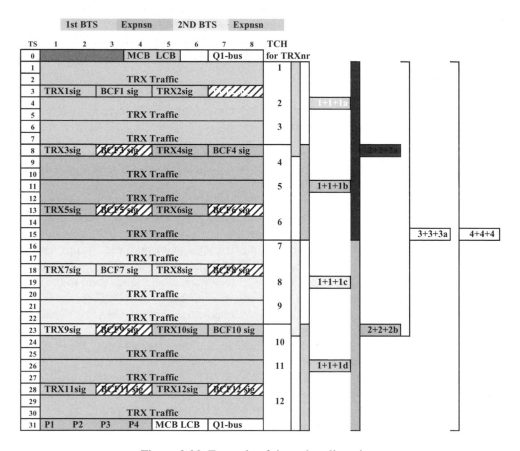

Figure 3.20 Example of time-slot allocation

which shows a $1 + 1 + 1$ configuration. The time slots for traffic and signalling for each TRX are allocated together. Time slots 1–3 contain the traffic for TRX_1 and signalling for TRX_1 and TRX_2. Subsequent time slots 4–7 are empty, because of an assumption that the capacity for cells will increase from one TRX to three TRXs. And same pattern is followed for other cells. This scheme can also be used for a condition where four base stations are using the same 2 Mbps signal. The synchronisation bits and management bits can be placed in either TS0 or TS31. These bits can also be placed in other time slots as well.

3.4.3 2 Mbps Planning

The routing of the 2 Mbps is part of time-slot allocation and detailed planning. This involves the planning of routes from where the 2 Mbps line will pass and where it will terminate. Planning engineers have to pass on these plans to the commissioning engineers, so they should be very carefully and clearly made. Apart from this, planning engineers during the course of 2 Mbps planning would calculate the number of 2 Mbps ports that will be used at the BSC.

The example shown in Figure 3.21 is for two sites that are connected between two BSCs. The sites are connected to each other by microwave radios. They are path-protected (i.e. in a loop) and thus 2 Mbps are terminating at cross-connect (here it is SXC or SDH cross-connect). Each site or base station is using one PCM and terminating at the ET ports at the BSC. Each of the PCM requires one port at the base station. In some cases, one PCM is shared between two or more base stations, as shown in Figure 3.20, but the number of ET ports required at the BSC would be still one. Transmission planning engineers should remember to reserve some ports for the 2 Mpbs connections between the BSC and MSC.

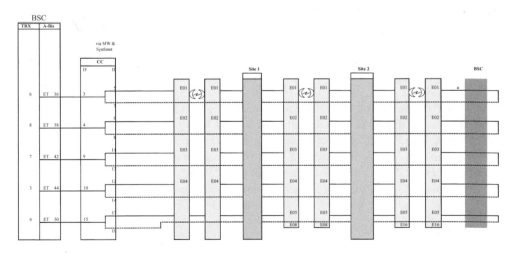

Figure 3.21 Example of 2 Mbps planning

In big upcoming networks, the 2 Mbps plans keep on changing. Planning engineers need to keep a track of how the 2 Mbps paths are being routed as this directly effects the ET port allocation plans and the complete utilisation of the PCM signals.

3.4.4 Synchronisation Planning

Another important aspect of detailed transmission planning is the synchronisation plans. Usually networks are designed using mixed technologies (SDH and PDH). In mixed networks that do not have proper synchronisation planning, 'clicks' are heard owing to bit errors in the data service (see Figure 3.22). SDH networks themselves do not require any synchronisation for proper operation. The reason is that the SDH payloads are easily transferred within the network while the frequency differences can be taken care of by a pointer mechanism. Pointers provide the mechanism for dynamically accommodating the variation in phase of the multiplex section into which they are to be multiplexed. Thus, when both PDH and SDH equipments are present in the network, degradation takes place due to the pointer movements. If the synchronisation planning is done properly, then these 'clicks' due to bit errors disappear.

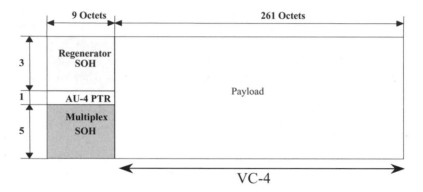

Figure 3.22 SDH frame for STM-1 interface. The frame size of an SDH frame is 260 (bytes) $\times$ 8 (bits) $\times$ 9 (rows) and the frame is transmitted $8000\,s^{-1}$, so the payload bit rate is 149.76 Mbps. The first column in the virtual container is the VC-4 path overhead (POH) and the header consists of regenerator and multiplex section overhead (SOH) and AU-4 pointer fields

Another reason for the need for synchronisation is that there are equipments/systems other than the SDH and PDH in the network, such as base stations and other transmission equipments. The internal clock of these equipments is not very accurate, so external synchronisation is needed for movement of 'click free' calls through the network. The required accuracy for the GSM networks is of the order of 10^{-8}. The quality of the clock is important in determining the quality level of synchronisation within the network.

There are three clock levels recognised by the ITU-T G.803:

- PRC (primary reference clock)

- slave clock (synchronisation supply unit or SSU)

- SEC (SDH equipment clock).

The primary reference clock usually provides the clock for the whole network, thereby acting as the master. The accuracy of the PRC is very high – usually of the order of 10^{-11}. The specification of suitable clocks which can act as the PRC is in ITU-R recommendation

G.811. As the PRC is the master, it is advisable to have a redundant source (i.e. two PRCs, with one clock in standby mode). Caesium clocks are a typical example and are (usually SSU) highly accurate, but their high cost may be unacceptable in small networks.

Slave clocks are used in conjunction with the master clock. These clocks are always locked into the clocks of higher accuracy (e.g. caesium) through the network. They are also known as 'refresher clocks' because they may be used in long chains to refresh the timings. They also have the capability to stand in for the master clock if the latter is lost temporarily. The accuracy of slave clocks is of the order of 10^{-8} based on ITU-T recommendation G.812. The major advantage of such clocks is their low cost, but lifetime may be limited.

The SDH equipment clock has an accuracy of the order of 10^{-4} based on ITU recommendation G.813. Usage of such clocks is recommended for SDH networks. Mixed networks (that have PDH equipments) should not use this clock for timing distribution as it may degrade the signal quality.

Synchronisation Planning Principles

As cellular networks are *mixed* networks, they generally follow the master–slave technique of distributing the clock signal. In this method, a higher-level clock synchronises the lower-level clock. The latter further distributes the timing signal as shown in Figure 3.23. The MSC receives the clock from the PRC and distributes it to the BSC. The BSC then further distributes it to the base stations, which further distribute the timing signal.

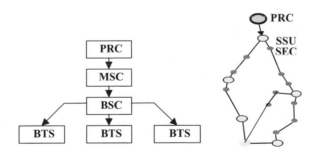

Figure 3.23 Clock distribution in a typical cellular network

Topology has an important place in synchronisation planning. Long chains of timing distribution should be avoided. There should not be more than 10 SSUs in the chain, and the number of SECs between two subsequent SSUs should not be more than 20. This implies that the maximum number of G.813 clocks in the synchronisation chain is 60, and the maximum number of G.812 clocks is 10.

If there are loops in the network, then engineers should make sure that there is no loop of the timing signal, to avoid a timing loop.

Synchronisation should be protected. There should be more than one source of the timing signal for each network element. The clocks should be derived from reliable sources such as SDH equipment, rather than from 2 Mbps signals from the leased lines whose accuracy is not known.

Implementation of the Synchronisation

The PRC is generally the source of the timing signal in a cellular network. The clock should be re-traceable back to the source. For loop protection, MCB and LCB bits in TS31 (or any other time slot) are used. If an MCB (master control bit) is used, this indicates whether or not the clock timing is based on the master, with '0' indicating that it is based on the timing sent by the master and '1' indicating that the timing is not based on the master. If synchronisation is based on the master–slave technique, then MCB '1' also indicates that there is some fault with the flow of the timing signal and the equipment is not receiving timing from the master. The LCB (loop control bit) would indicate '0' when there is the possibility of a timing loop, which means that if there is a loop topology in the network, the LCB would be '0' indication that the timing loop is not there (based on the principle of synchronisation planning explained above). As every element should have more than one clock source, so the LCB indicates from where the equipment should take its timing signal. So:

- MCB: 0 – signal is based on the master clock

- MCB: 1 – signal is based on some other clock (e.g. internal)

- LCB: 0 – synchronise slave from master, no possibility of timing loop

- LCB: 1 – do not synchronise slave from master (e.g. there may be possibility of a timing loop).

3.4.5 Transmission Network Management Planning

Network management systems (NMS) can be of many types. They may be able to manage the whole network or may be specifically for management of the transmission element. Generic applications of network management systems include network element (and configuration) management, security management, alarm management, performance and fault management.

Network element (NE) and configuration management functions may include controlling the NE, collecting and providing the data to the NEs. This function also makes integration of the NE to the network possible. The alarm management function enables the user to collect the alarms from the NEs. These alarms indicate the status of these NEs. The fault management function detects failures and schedules the correction of these faults, apart from testing and bringing back the faulty NE back into working condition. The generation of the performance behavioural reports of the network and its elements is done by the performance management function of the network. And finally, prevention and detection of misuse of the network from breaches of security is taken care of by the security management function of the NMS.

Network management functionality is based on master–slave protocols, and planning of the network management system starts with the choosing of the master. The NMS can act as the master and the NEs as the slave. Sometimes, network elements such as the base station controller or the base stations can also be the master. Once the master is defined, the management bus and its transfer method is decided. Then the next step is to decide the parameters for each of the network elements that the master will control.

Each master that is chosen has its own capability of managing a number of network elements. The capacity may vary from managing just a few NEs to hundreds of NEs. Apart from that, each master has its own speed of managing the elements; i.e. the speed of the management buses may vary in the network. These management buses may have a speed of 1200 bps, 9600 bps or even higher.

Definition of the management bus includes deciding the NEs that will be engaged by a particular network element. Also, the topology of the management bus and the protection scheme that is associated with it are decided. One thing to remember is that the higher the number of elements associated with the management bus, the slower will be the speed of the bus.

Network elements can be managed by attaching the slave to masters through a cable. Other techniques involve sending the managements bits through the PCM signal from TS0 to TS31 and through the auxiliary data channels of frame overhead (of for example radio frames).

Engineers need to define parameters such as the addresses of the NEs along with their identifications, and the speed to the management buses. These parameters depend upon the type of management system that is being used in the network and its capability to handle the amount and types of network elements. One typical example of parameter settings is shown in Figure 3.24. There may be more parameters than shown depending on what type of master is being used and how the management channels are being transferred and controlled.

Group	Setting	Value/Operation	Comments
Identification	Equipment Type	MWR	
	Equipment Number	2233	
	equipment name station ID	according the transmission plan / station identification	
Service options	baud rate	1200 bit/s	Default: 9600 bit/s
	address	1 Ö 4000 (Individual address)	Default: 4095
Equipment settings	Mgmt Ch. Loop protect.	OFF	Default: OFF
	RX Mgmt Channel	Disconnected	Default: Normal

Figure 3.24 Example of parameter setting for microwave radio

3.5 TRANSMISSION NETWORK OPTIMISATION

3.5.1 Basics of Transmission Network Optimisation

In the past, the term 'network optimisation' meant solely radio network optimisation. Transmission planning itself was considered to be a line-of-sight job, and any problems in the network coverage capacity or quality were attributed to the radio part of the network. With more and more studies being conducted on mobile networks, it was found that transmission and the core network play an equally important role in contributing to a highly efficient network. Moreover, some networks that came during the evolutionary years of GSM were designed in a hurry, without much consideration for future capacity requirements and the knowledge that the quality of transmission network would affect the quality of the whole network.

In Chapter 2 we saw that radio network optimisation is a process that starts almost simultaneously with the radio planning process. This is not the case with transmission network optimisation. The main reason for this is the static nature of the transmission network. The transmission network, once it is up and running, should not be touched until and unless a problem in identified and verified. For example, if one microwave link near the BSC in a chain is 'down' for few minutes, the network traffic and hence revenue generated from all the sites getting connected to that BSC would be lost. For this very reason, the transmission networks are *not touched* during the network optimisation.

The process of transmission network optimisation is shown in Figure 3.25. When it has been identified that the transmission network needs to be optimised, the target areas for optimisation are defined. The process starts with the making of plans based on the data available and the targets to be achieved. These plans are then assessed and optimised, to achieve a balance between costs, quality and the time frame for the process. The process of optimisation generates some plans, and these plans are then analysed. Again, based on the requirements and the costs, a master plan is prepared. This master plan is then implemented. Unlike in the radio network, where optimisation is an ongoing process, there is a gap period before the next optimisation cycle can begin in the transmission network.

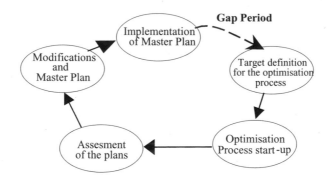

Figure 3.25 Cellular transmission network optimisation cycle

3.5.2 Transmission Network Optimisation Process

Process Definition

There are two major reasons for transmission network optimisation: either capacity is less than desired, or the quality is not up to standard (see Figure 3.26). Once the main problem has been identified, the key performance indicators (KPIs) should be decided upon. The main KPIs in a transmission network are related to quality of the microwave links such as ESR and SESR. Apart from this, some parameters like slip frequency can be observed during a synchronisation study.

Capacity Analysis

During the initial network *rollout*, the benchmarks are usually on how many sites are *on air*. For going on air, the base stations need to be connected to the BSC. When the deadlines are fast approaching, the usual way to connect these sites is through leased lines, so topology

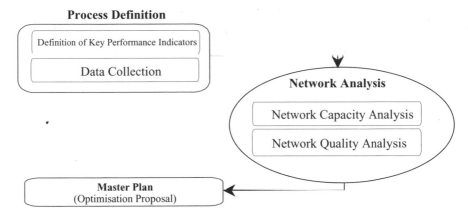

Figure 3.26 Cellular transmission network optimisation process

design, its effects and benefits may be neglected. Temporary solutions may result in under-utilisation of the media that is connecting the sites. Another major factor is congestion in the network that may be due to the increase in the number of mobile users. This may lead to an increase in the number of TRXs at each site, thereby forcing the transmission planning engineers to come up with solutions to cater for the traffic increase without increasing the number of base stations. Another reason could be the need to upgrade the network technologically, e.g. 2G to 2.5G or 3G.

The capacity optimisation process ends up with a twofold solution. It provides answers as to how to use the existing capacity effectively, and how to increase the capacity with minimal disturbance to the network functionality and revenue inflow. Apart from this, the network will be more flexible and reliable.

How should capacity analysis be done? Engineers first collect regional data. Apart from having site information (site location, coordinates, height, etc.), information collected should include site capacity (present as well as forecasted), 2 Mbps routing plans, and media information (e.g. leased line, microwave radios, optical cables, etc.). Coverage plans will also help if new sites are being introduced.

From this information, transmission planning engineers should calculate the existing and forecast capacity of each link. The next step is to identify the media. Base stations that have no line-of-site (LOS) to another base station or BSC might be using leased lines/optical cable to get directly connected to the BSC (such sites are usually small in number). The capacity utilisation of all other sites should be assessed. Is there a possibility to replace the leased lines with microwave radio? With microwave radio, the sites can be connected to each other (if LOS exists) and the capacity increased many times. Leased lines are usually PCM type while the microwave links are usually 2E1 to 16E1 capacity. This not only increases the spare capacity of the network, but also increases the chances of creating topologies thereby increasing the protection level.

One such analysis is shown in Table 3.2. There, eight sites are analysed in a region. Each site is analysed and the topology change is proposed along with the change in the media from leased line to microwave radio. Though for some sites (e.g. site 74) there is no improvement in the capacity, the overall capacity of the system improves by over 160% This example considers just a few sites, but when capacity optimisation takes place, minor changes in topology may lead to a tremendous improvement in capacity.

Table 3.2   Example of capacity analysis

S.No.	Site ID	Connectivity before Optimisation	Capacity (2 Mbps/site) (Present)	Connectivity after Optimisation	Capacity (2 Mbit/site) (After)	Capacity improvement (%)
1	82	82-50; 50-06	0.25	82-50; 50-06	0.50	100.00
2	83	83-82; 30	0.25	64-83; 64-06	2.00	700.00
3	87	70-87; 70-69; 25	0.33	70-87; 70-69; 69-06	0.80	140.00
4	2	61	1.00	61	1.00	0.00
5	35	35-82; 31	0.25	35-82; 50-06	0.50	100.00
6	36	4	1.00	36-06	4.00	300.00
7	73	73-76; 38	0.50	73-06	1.33	166.67
8	74	74-82; 50-82; 50-06	0.50	74-82; 50-82; 50-06	0.50	0.00
		Av. capacity per site (before)	0.11	Av. capacity per site (after)	0.28	160.41

If new sites are to be added at the hot spots during/after the optimisation process, then the capacity increase in particular links/region should be considered apart from considering the impact of the new sites.

The results of the capacity analysis should be very well understood and its effect on quality and costs analysed before making recommendation for changes in the master plan for the implementation of these recommendations.

Quality Analysis

Network quality degradation can have several causes. These include reduced microwave link performance, interference, and synchronisation-related problems.

The most common reason for degradation in microwave links is varying and unpredictable propagation conditions. Earlier in this chapter, conditions such as fading (flat, multipath and k-fading), layering, ducting, attenuation due to rain, etc., were discussed. However, when the signal fades and stays that way, it is essential to look for the cause(s) and remedy the situation. The quality of microwave links is measured in terms of error performance and availability. When the quality targets have been checked and redefined, the process of data collection starts.

This data generally includes planning data (topology, hop lengths, planned link budget calculations, frequency plans etc . . .) and link performance data that should contain the performance data at the time of link commissioning and during the problem period. Also, data collected during different periods of the year and at different times of the day would be of help. Mobile call statistics during the fading and non-fading phases would enable a better understanding of the correlation, if any, between call drops and fading.

Analysis of the data should begin with recalculation of the link budgets so as to confirm that there were no errors in the initial design. The data of these links monitored at the time of commissioning and during the optimisation phase should be analysed against the backdrop of the design data in the master plans (made before the network launch). One such example of link data calculated and monitored during the optimisation phase is shown in Figure 3.27. The unavailability figures of the four links in a region are calculated and the worst case is analysed by using the data recorder for a period of 24-hours.

Once it has been detected that the received level is below the desired level, path profile analysis should be done again in case the path lengths are more than a few kilometres. Also, error performance statistics such as the ES (error seconds) and SES (severe error seconds) should be analysed. These statistics can be collected using the NMS, as shown in Table 3.3.

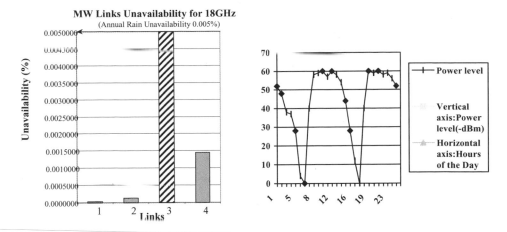

Figure 3.27 Unavailability graph for four links and power level observation of an MW link for a 24-hour period

Table 3.3   Statistics from NMS

	Avail_time	Degraded_min	Err_sec	Err_sec_severe
1/10/1997	86304	0	0	0
1/11/1997	86640	0	0	0
1/12/1997	83059	0	0	0
1/13/1997	83640	0	0	0
1/14/1997	63598	0	0	0
1/15/1997	63766	0	0	0

In general, the dropping of calls in a mobile network is attributed to the radio network. In fact, in some cases, it may be due to problems with interference and synchronisation. The traffic lost in the form of dropped calls can be as high as a few thousand minutes for a small 100-site network, as shown in Figure 3.28. This results in a huge revenue loss.

With microwave transmission, the possibility becomes higher for interference being the cause of 'trouble'. The trouble-shooting process starts with a search for the cause of dropped power levels at a hop (or hops in a region). The interference problem analysis in the links or region may include the study of existing frequency plans, or a practical study consisting of frequency scanning. The interference could be due to an internal or external source; an internal source being easier to deal with. Once the interference source has been identified, a solution has to be applied. This may include the changing the frequencies, reducing the power level of the interferer, antenna discrimination, etc.

Synchronisation can be one of the reasons behind degraded network quality. This problem usually persists in mixed networks. The data collected would include mainly the synchronisation plans (on whose basis the implementation was done), and NMS statistics on whether the slip frequency limit is exceeded. The primary source of error could be in the clock flow (i.e. MSC to BSC to BTS, etc.). In many cases it is discovered that the accuracy of the PRC is unsatisfactory (against given standards for mobile networks)

Typical outputs of link budget calculations, frequency plans and interference are shown in Tables 4 and 5, and Figure 3.29.

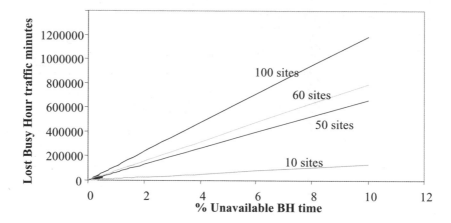

Figure 3.28 Lost traffic minutes versus percentage unavailable time per year

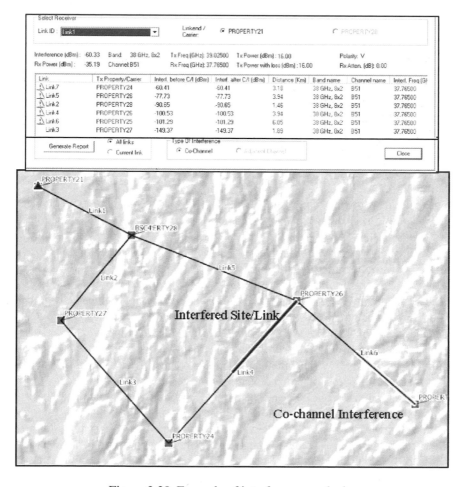

Figure 3.29 Example of interference analysis

Table 3.4 Link Data for the Test Network

Link ID	Link Type	Capacity	Length (km)	LOS Status	Calculation Method	pL Value	C/K Factor	Climate Region/Rate	% Time Rainfall	Band	Channel	Centre Frequency (GHz)	Bandwidth (MHz)	Frequency Designation	Latitude	Longitude	MTTR	Antenna Size (m)	Antenna Height (m)
Link1	Default Microwave	16.0Mb/s	1.46	OK	ITU-R P.530-10	10.0	0.00042951	E	0.0100	38 GHz, 8 × 2	B51	39.0250	14.0000	High	60°21'58.76"N	024°38'5.03"E	6.000000	0.60	4.00
										38 GHz, 8 × 2	B51	37.7650	14.0000	Low	60°21'38.59"N	024°39'31.16"E	6.000000	0.60	6.00
Link2	Default Microwave	16.0Mb/s	1.54	OK	ITU-R P.530-10	10.0	0.00043256	E	0.0100	38 GHz, 8 × 2	B52	39.0390	14.0000	High	60°21'38.59"N	024°39'31.16"E	6.000000	0.60	11.00
										38 GHz, 8 × 2	B52	37.7790	14.0000	Low	60°20'58.88"N	024°38'30.64"E	6.000000	0.60	13.00
Link3	Default Microwave	16.0Mb/s	2.26	OK	ITU-R P.530-10	10.0	0.00042909	E	0.0100	38 GHz, 8 × 2	B51	39.0250	14.0000	High	60°20'5.75"N	024°40'11.93"E	6.000000	0.60	42.00
										38 GHz, 8 × 2	B51	37.7650	14.0000	Low	60°20'58.88"N	024°38'30.64"E	6.000000	0.60	44.00
Link4	Default Microwave	16.0Mb/s	2.68	OK	ITU-R P.530-10	10.0	0.00042625	E	0.0100	38 GHz, 8 × 2	B52	39.0390	14.0000	High	60°21'12.28"N	024°42'4.14"E	6.000000	0.60	25.00
										38 GHz, 8 × 2	B52	37.7790	14.0000	Low	60°20'5.75"N	024°40'11.93"E	6.000000	0.60	25.00
Link5	Default Microwave	16.0Mb/s	2.48	OK	ITU-R P.530-10	10.0	0.00043086	E	0.0100	38 GHz, 8 × 2	B52	39.0390	14.0000	High	60°21'38.59"N	024°39'31.16"E	6.000000	0.60	26.00
										38 GHz, 8 × 2	B52	37.7790	14.0000	Low	60°21'12.28"N	024°42'4.14"E	6.000000	0.60	27.00
Link6	Default Microwave	16.0Mb/s	2.18	OK	ITU-R P.530-10	10.0	0.00043869	E	0.0100	38 GHz, 8 × 2	B52	39.0350	14.0000	High	60°21'12.28"N	024°42'4.14"E	6.000000	0.60	25.00
										38 GHz, 8 × 2	B52	37.7790	14.0000	Low	60°20'27.71"N	024°43'54.47"E	6.000000	0.60	25.00

Table 3.5 Detail link performance for the test network

Link ID	Freespace Loss (dB)	Atmospheric Absorbtion (dB)	Obstruction Loss (dB)	Total Loss (dB)	Rx Level (dBm)	Threshold Value (dBm)	Threshold Degradation (dB)	Composite Fade Margin (dBm)	Flat Fade Margin (dB)	Flat Fade Margin After Interference (dB)	Req. FM Against Rain (dBm)	Interference Margin (dB)	Dispersive Fade Margin (dB)	Flat Outage (PnS) (%)	Selective Outage (Ps) (%)	Red. Of X Polar Discrimination (%)	Total Worst Month Outage (Pt, w/o Div) (%)
Link 1	127.4243	0.1633	0.0000	138.5876	−35.1876	−85.0000	0.0000	48.8959	49.8124	49.8124	7.0421	0.0000	56.1024	0.0000000	0.0000000	0.0000000	0.0000000
	127.4243	0.1633	0.0000	138.5876	−35.1876	−85.0000	0.0000	48.8959	49.8124	49.8124	7.0421	0.0000	56.1024	0.0000000	0.0000000	0.0000000	0.0000000
Link 2	127.8877	0.1723	0.0000	129.0600	−25.6600	−85.0000	0.0000	54.4160	59.3400	59.3400	7.4068	0.0000	56.1024	0.0000000	0.0000000	0.0000000	0.0000000
	127.8877	0.1723	0.0000	129.0600	−25.6600	−85.0000	0.0000	54.4160	59.3400	59.3400	7.4068	0.0000	56.1024	0.0000000	0.0000000	0.0000000	0.0000000
Link 3	131.2242	0.2529	0.0000	132.4770	−29.0770	−85.0000	0.0000	53.0015	55.9230	55.9230	10.5880	0.0000	56.1024	0.0000001	0.0000002	0.0000000	0.0000000
	131.2242	0.2529	0.0000	132.4770	−29.0770	−85.0000	0.0000	53.0015	55.9230	55.9230	10.5880	0.0000	56.1024	0.0000001	0.0000002	0.0000000	0.0000000
Link 4	132.7115	0.3003	0.0000	134.0118	−30.6118	−85.0000	0.0000	52.1510	54.3882	54.3882	12.3770	0.0000	56.1024	0.0000001	0.0000004	0.0000000	0.0000000
	132.7115	0.3003	0.0000	134.0118	−30.6118	−85.0000	0.0000	52.1510	54.3882	54.3882	12.3770	0.0000	56.1024	0.0000001	0.0000004	0.0000000	0.0000000
Link 5	132.0359	0.2778	0.0000	133.3137	−29.9137	−85.0000	0.0000	52.5544	55.0863	55.0863	11.5339	0.0000	56.1024	0.0000001	0.0000002	0.0000000	0.0000000
	132.0359	0.2778	0.0000	133.3137	−29.9137	−85.0000	0.0000	52.5544	55.0863	55.0863	11.5339	0.0000	56.1024	0.0000001	0.0000002	0.0000000	0.0000000
Link 6	130.9188	0.2443	0.0000	132.1630	−28.7630	−85.0000	0.0000	53.1589	56.2370	56.2370	10.2530	0.0000	56.1024	0.0000000	0.0000001	0.0000000	0.0000000
	130.9188	0.2443	0.0000	132.1630	−28.7630	−85.0000	0.0000	53.1589	56.2370	56.2370	10.2530	0.0000	56.1024	0.0000000	0.0000001	0.0000000	0.0000000

4

Core Network Planning and Optimisation

4.1 BASICS OF CORE NETWORK PLANNING

4.1.1 The Scope of Core Network Planning

The core network in GSM is basically the circuit core. In GPRS, EDGE and UMTS, it has two components, the circuit core and the packet core, which are responsible for voice and data respectively. In GSM, core or circuit-core network planning is also known as switch planning, because it is mainly related to the mobile switching centre (MSC), also called 'Switch'.

Core network planning in GSM consists of network elements such as MSC, VLR, HLR, AC and EIR.

4.1.2 Elements of the Core Network

MSC/VLR

The MSC is the core element of the network sub-system. It is responsible for the switching of subscriber calls. It is also responsible for traffic management, paging, and the collection of charging-related information. Apart from this, the MSC also acts as an interface between the network and the networks of other operators and PSTN networks. Another important element that is hosted in an MSC is the visitor location register (VLR). The VLR contains information on the subscribers who are being handled by the MSC at a given moment. Owing to the nature of its functionality, the VLR participates in call processing and mobility management functions of the network such as the use of temporary mobile identification, IMSI attach/detach, etc., apart from location registration of the mobile subscriber. The

Fundamentals of Cellular Network Planning & Optimisation A.R. Mishra.
© 2004 John Wiley & Sons, Ltd. ISBN: 0-470-86267-X

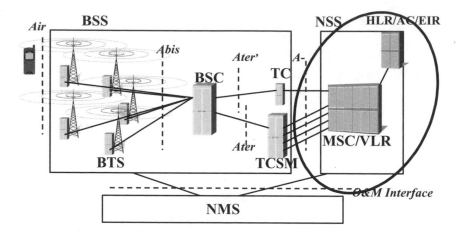

Figure 4.1 Scope of core network planning

VLR is also involved in security functions such as IMEI (International Mobile Equipment Identification) checking.

HLR/EIR/AC

The home location register (HLR) is responsible for keeping all permanent information regarding subscribers and their equipment. It is also responsible for creating, modifying, deleting or managing these data. Apart from this, the HLR also participates in call processing functions such as routing, supporting the incoming call barring service/unconditional call forwarding service, etc.

The HLR consists of an EIR and an AC. The EIR (equipment identification register) contains databases for IMEI. The EIR is the place where mobiles that are missing or stolen are registered. Actually the EIR contains three lists: white, grey and black. The white list contains IMEI for all the authentic mobiles; the grey list has data for faulty mobiles; and the black list contains information about stolen/missing mobiles. Thus, when a call is being made, the number is cross-checked at the EIR. If it is found in the white list, the call goes ahead, but if the number is found in the black list, the call may be blocked.

The AC (authentication centre) is mainly responsible for security aspects. As the name suggests, it manages the security information for authentication of subscribers. This information is usually requested by the VLR and, in coordination with the EIR, it prevents an unauthenticated card from accessing the network.

4.2 CORE NETWORK PLANNING PROCESS

The planning process for the core network is slightly different from the radio or transmission network planning processes. It consists of two parts: switch network planning and signalling network planning. In the radio and transmission network planning processes, environmental factors and site selection processes assume great importance. However, these two processes are missing from core network planning, thereby reducing the process to three main steps (network analysis, network dimensioning and detailed planning) before drawing the final core network plan. These steps are shown in Figure 4.2.

Network analysis and dimensioning are explained separately for switch and signalling network planning.

Figure 4.2 Core network planning process

4.2.1 Network Analysis

The number of switches (MSCs) in any network is far less than the number of sites or links. Some networks even have only one switch, which means that they cannot be changed on a 'daily basis' as that would have an impact on the performance of the whole network. Switch planners have to take forecasts into account because switches are expected to cater for the expected traffic rise for a few years before their number is increased.

Network analysis requires data ranging from subscriber information to demographic information. The major dataset constitutes existing network data, the existing service plan, subscriber base data, topographical data, and traffic data (both existing and forecast). Apart from these, some radio network planning data and/or transmission planning data might also be needed.

Existing network data gives the core-planning engineer information about the number of sites and the traffic it is expected to generate. The total traffic within a network usually consists of traffic that is generated within the network and traffic that was generated outside the network (e.g. calls being made to mobile subscribers from fixed lines). This traffic distribution is very helpful during both the planning and optimisation phases of the core network.

The next thing to find out about is the services to be delivered to mobile subscribers, and the traffic it is expected to generate. Apart from voice, this generally includes value-added services such as SMS, MMS, Internet, etc.

Subscriber base data is one of the major inputs for switch planning. It includes the existing subscribers or subscribers expected at the time of network launch, the expected traffic that will be generated by these subscribers, forecast subscribers and forecast traffic. The forecast may be in phases that may depend upon expansion of the network or the subscriber base. The information related to the different regions (urban, rural, hot spots, etc.) and the type of subscriber base (business users, residential users, etc.) will give a fair idea of the expected traffic. Other useful information comes from a numbering study, including roaming numbers, IMSI numbers, signalling point codes (SPC), emergency numbers, allotted subscriber numbers, etc.

4.2.2 Network Dimensioning

All the information collected during the network analysis phase is required in the dimensioning phase. So, what is the output of the dimensioning exercise? It is the number of nodes that are required for handling the subscribers (and traffic) efficiently for a longer period of time. Here is a brief overview of the major outputs of dimensioning of the switch network:

- expected traffic generated in the network

- the number of switches required to handle subscribers and traffic (both present and forecast)

- the most efficient location of the switches in the network

- how the switches will be connected to each other (i.e. transmission plan for the switches)

- the most efficient way to route the traffic.

The traffic calculation is perhaps the most important aspect of switch planning. It has to be as accurate as possible and all the factors discussed in the previous section contribute to this. One such example is shown in Figures 4.2(a) and (b). When there is more than one switch or more than one external network (i.e. PSTN, other operator, etc.), the generation table becomes bigger and complicated as traffic flow from each of the switches to/from the other switches and external network need to be taken into account. Of all the traffic existing in a cellular network, some percentage is generated and terminated in the same network, while some traffic terminates in other networks and vice versa.

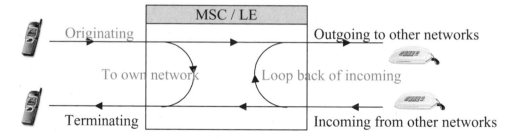

Figure 4.2(a) Traffic flow

From	MSC	External Network
MSC	X	800
External Network	1600	X

Figure 4.2(b) Traffic generation (in erlangs)

There follows a simple example. Usually there would be multiple switches, and in such cases the traffic scenario (also known as a traffic matrix) becomes more complicated; inputs like the number of subscribers calling from mobile to mobile, mobile to fixed and fixed to mobile are required.

Example 1: Basics of a Traffic Calculation
Consider, then, a simple core network with one switch. The present traffic is due to its own subscribers and traffic coming in from external sources (or local exchange). This is the

scenario represented by Figure 4.2(a). Before the dimensioning starts, few parameters need to be understood and defined:

- *Subscribers originating (SO). This is the traffic originated by subscribers of the network. Generally, this input comes from the network (or network operator). Typical values can be 65% for the switch and/or 50% for the LE.*

- *Subscribers terminating (ST): This is the traffic that is being terminated in the same mobile network. This value can be calculated as ST = 100% minus SO.*

- *Own network (terminating) traffic: This is the traffic that is originated in the network and terminated there also. It is usually a product of subscribers in the network and SO.*

- *Loop back of incoming traffic: This is the traffic originating from the external networks and routed back to them.*

- *Outgoing and incoming traffic: Traffic going outside the mobile network to an external network is outgoing traffic, while traffic coming from external networks is incoming traffic. These can be calculated as follows:*

  Incoming = (ST – own network traffic)/(1 – loop back of incoming traffic)

  Outgoing = SO – own network traffic + loop back of incoming trafffic.

- *Voice-mail system and interactive voice response (IVR) traffic: Traffic related to voice-mail is of two types: forwarded and listen. If the traffic is diverted to voice-mail when the subscriber is not available, it is known as VMS forwarded traffic, while traffic generated to listen to voice messages is the VMS listen traffic. IVR is the traffic generated through interactive responses, when a recorded voice responds rather than a subscriber (e.g. 'Press 1 for Hindi, press 2 for English . . . ').*

- *Inter-switch traffic: This is calculated on the basis of the incoming and outgoing traffic. Generally tools are used for this. An example is shown later in Figure 4.6.*

Another important parameter related to subscriber behaviour is the 'calling and moving interest'. Calling interest indicates the distance to which calls are made (i.e. long distance or short distance), as shown in Figure 4.3. This figure is useful for inter-switch modelling when

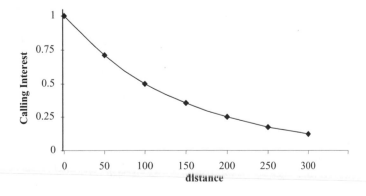

Figure 4.3 Calling interest

calculating the number of subscriber-originated calls. Moving interest is another parameter that defines the subscriber behaviour of moving within the network. The external calls would then be diverted via the shorted path using the HLR-call enquiry feature.

When the number of subscribers has been determined, the number of switches can be calculated. The subscriber number directly affects the visitor location register (VLR). Thus:

Number of switches = number of subscribers/VLR (or HLR) capacity.

When the number of switches has been determined, the next step to locate them. If there is only one switch, the decision is quite easy: it is generally located near the mobile headquarters or at a place having easy accessibility. In this case the expected traffic is also low, so no routing plan is required as all the traffic goes to only one switch. However, the scenario changes quite a bit when there is more than one switch. The main idea is to keep the switch locations in areas of high subscriber density, thereby saving on transmission costs.

Next, the traffic route needs to be defined. Routing may be done in two ways: partly mesh or completely mesh. Assume that there are four switches. The traffic can be routed as shown as in Figure 4.4(a), where all the MSCs are connected in cyclic fashion with some protection afforded by a diagonal connection. In Figure 4.4(b), all the MSCs are interconnected, so traffic from each flows into the rest of the MSCs. Though the former technique is simple, it results in a higher amount of transit traffic. Obviously the traffic in the complete mesh is more protected. This is a simple case. If the number of switches increases then the mesh technique becomes more complicated and routing of all the traffic to all the switches becomes a difficult task. In such cases, another dimension is added to routing, i.e. transit switches are added.

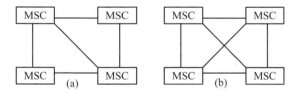

Figure 4.4 Interswitch connection

If a transit switch (TS) were added to the network, it would look like Figure 4.5. All the traffic, both internally and externally generated, will be routed through the transit switch. As the transit switch has great importance in this arrangement, it is often advised to have a second redundant transit switch. If the cellular network is spread over a very large region and has many switches, then the number of transit switches increases also. Although the MSCs

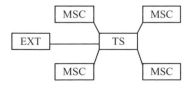

Figure 4.5 Switch network with transit switch

are connected to the nearest transit switches, the network topology between the transits switches themselves is fully meshed.

Next the routing plan is devised the flow of traffic between switches. One single route must not be overloaded with traffic; rather, the traffic needs to be evenly divided between various routes in the best possible way. The routing plan gives the direct path that is taken by traffic and, in case of failure of the primary route, the secondary path taken. The secondary path choice is dependent on the traffic it is already carrying. Thus, even in the initial days of the network, spare capacity planning is required so as to route calls through the secondary paths when the primary ones are blocked.

An example of traffic dimensioning is shown in Figure 4.6. There are four MSCs and an external network. The number of subscribers handled by each MSC is 50,000. With the traffic generated by each subscriber being 27 mErl (milli-Erlangs), the total traffic generated by each MSC is 1350 Erl. The percentage of calls originating within their own switch is 60%, of which 40% terminate in the same network. Calling interest and moving interest are 215 km and 50 km respectively.

Input							
	MSC1	MSC2	MSC3	MSC4			
X	0	0	200	100		call int	215
Y	0	0	0	100		move int	50
subscribers	50000	50000	50000	50000			
tr/subs (mErl)	27	27	27	27			
total traffic	1350	1350	1350	1350			
subs originating	60%	60%	60%	70%			
subs terminating	40%	40%	40%	30%			
to own network	20%	20%	20%	20%			
to own NW of total	12%	12%	12%	14%			
to own NW of total	162	162	162	189			
loopback	5%	5%	5%	5%			
initial outg ISW	110	110	93	136			
initial outg ISW%	8%	8%	7%	10%			

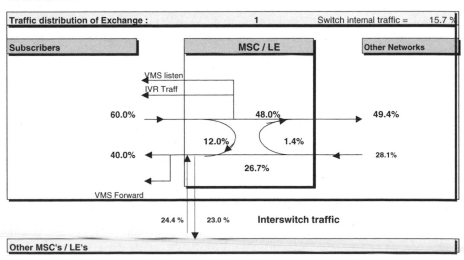

Figure 4.6 Example of traffic dimensioning

4.3 BASICS OF SIGNALLING

4.3.1 Signalling Points

As mentioned in Chapter 1, signalling in the NSS network is SS7. Signalling can be transferred on 64 kbps time slots in PCM. Any point that is capable of sending or receiving the signalling is known as a signalling point (SP). Signalling traffic passes through a signalling transfer point (STP) and reaches its intended destination, known as the signalling end point (SEP). Thus the SP and the SEP are the same. SPs are connected to each other via PCM links. A SP can also act as a signalling control point (SCP). Through this point, access to advanced services can be made, such as freephone numbers.

Signalling networks can be of different types. Signalling network indicators give the information on the type of signalling network used. In GSM networks, as SS7 is used for signalling, four different types of networks are possible: NA0, NA1, IN0 and IN1. The first two are for national networks and the last two for international networks. A SP can support one or more network addresses. It can also support all four networks, but in that case it would have four different SPCs, one for each network.

4.3.2 Signalling Links

A signalling link is a logical connection between two SPs. SPs are connected using PCM links (also known as PCM circuits). The purpose of the signalling link is to carry the messages of higher layers. Thus, a signalling link has an ability to define the start and end of a frame structure, locate the frame for initial alignment, maintain the signalling link, detect errors, etc. Usually two signalling links are used between two SPs, for protection purposes. These two signalling links are sent on two physically separate PCM links, constituting a signalling link (SL) set as shown in Figure 4.7. The maximum number of signalling links that can constitute a signalling link set is also limited, e.g. 16. These signalling links are usually assigned priorities. If there is a failure, the highest priority link carries the traffic of the failed signalling link.

For protection purposes, different routes are designated to carry the signalling from one destination point to another. These destination points are known as destination point codes (DPC). There is also a limit to the number of routes that can be assigned between two DPCs. However, there should be a minimum of two routes for protection purposes.

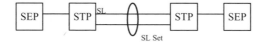

Figure 4.7 SEP, STP, SL and SL set

4.3.3 Signalling Network Dimensioning

Signalling is transferred on 64 kbps time slots on the PCM. High-capacity links use more capacity (i.e. more than one time slot). But as traffic increases, delays will take place in delivering messages. There are land-, equipment- and satellite-based signalling links. The acceptable delays for each of these can be 0.2 Erl, 0.4 Erl and 0.6 Erl respectively. Of these three values. 0.4 Erl is the performance requirement of the equipment nodes and 0.6 Erl is used for satellite connections. The main aspect to be remembered during dimensioning is to not exceed the recommended traffic load.

The process of dimensioning of the signalling network is as follows.

- dimension the end-to-end traffic (based on capacity assessment)

- route the traffic

- work out the transmission required.

The end-to-end traffic calculation is perhaps the most tedious as it is the whole basis of signalling dimensioning and planning. The main parameter inputs required for this include call-related parameters (i.e. those relating to successful calls). Short calls are essentially considered in core network planning because they generate large signalling traffic. They also include parameters related to the ISDN user part and telephony user part (i.e. TUP/ISUP). Another aspect to be considered is the traffic generated by messages (mobile application part (MAP) parameters). Routing-related parameters such as signalling link utilisation, signalling link set, etc., form another set of parameters to be considered for dimensioning. TUP/ISUP are call-related signalling while MAP is non-circuit-related signalling. Then there is signalling related to IN, which is the subject of the next section.

4.4 THE INTELLIGENT NETWORK (IN)

The 'intelligent network' concept in the core network permits services to be introduced in a network in a more cost-effective way. It also allows these services to be managed and controlled more effectively. This is done by implanting these services in a common/standard database instead of implanting them in each network element. This implementation makes the network more 'intelligent'. Pre-paid SIM cards, originating-call screening, reverse charging and freephone (1-800-numbers) are examples of services provided by an IN network.

For planning such a network, inputs such as the number of initial and forecasted subscribers, traffic behaviour, network topology, etc., are used. During the switch and signalling planning, IN network requirements must be taken into account as well. IN network analysis sets the performance objectives and connection requirements based on the traffic estimates. This is done for each of the IN module/services, and then finally the total traffic demand from all the IN services offered by the IN platform is calculated. Based on these inputs, IN application protocol (INAP) requirements and the subsequent IN platform (on which IN services run) are determined. Then some of the switches are upgraded by the addition of IN-capable platforms, and traffic from the remainder is routed through these switches.

The impact on the core network will be in terms of the capacity used by these IN services, apart from the fact that IN services generate their own signalling, thus increasing the total signalling in the core network.

One of the most important elements in an IN network is the IP (an intelligent peripheral). This is basically a stand-alone processor that provides the additional services for which IN is implemented in a network. IP functions include IVR, DTMF (dual-tone multi-frequency) translation, speech recognition, providing access to the signalling networks, etc.

Taking into account all the above factors, the number of signalling links can be calculated. Usually, network planning tools are used for this (see Appendix A). The output of signalling network dimensioning is the required number of signalling links. Once this is done, the number of ET ports can be calculated.

4.5   FAILURE ANALYSIS AND PROTECTION

As noted earlier in this chapter, the number of switches is fewer than the number of base stations or base station controllers, and in some cases there is only one switch in the network. A minor failure in the core network may lead to a large part of the radio network becoming non-functional, leading to a huge revenue loss. Protection of the network therefore becomes an inseparable part of core network planning. For designing (or assigning) protection, it is important to know the main failures that may take place in a network. These are:

- site failure

- equipment/node failure.

A site rarely fails. If it does, this is usually due to conditions that are beyond human control – flooding, earthquake, fire, etc. Although natural calamities cannot be prevented or their effects predicted, some steps should be taken to minimize the likely damage they cause. This may include choosing a site at a higher altitude in regions where floods are common, avoiding buildings or locations that do not have proper fire detection and prevention equipments, and adequate cooling arrangements to prevent over-heating of network elements.

Apart from site failures, small failures arising from power or transmission problems may lead to temporary failure of the equipment or node. One MSC and one HLR handle respectively about 0.5 million and one million subscribers (these values depend on the equipment manufacturer and individual equipments capacities). Thus, if one MSC or one HLR fails completely, revenue is lost from 0.5 to 1 million subscribers. Using redundant MSC/HLRs is one way to reduce the lost traffic.

The situation in networks having just one switch is quite severe in cases of failure. In networks having more than one switch, failure can be alleviated. One of the methods is to move the base stations from a failed MSC to another 'live' or redundant MSC. The second 'live' MSC should have enough capacity to handle the additional traffic.

In traditional networks, the HLR data were transferred to the redundant HLR using DAT tapes. These days, real-time updating of subscriber data takes place between the 'live' and redundant HLR.

Sometimes only part of some equipment may fail, but even slight damage/malfunctioning can lead to link or route failures. Typical examples of this are a damaged or disconnected PCM cable or ET port failure. There may be cases when route failure takes place, and existing routes experience higher loads leading to bigger delays. To cover these circumstances, it is recommended to define more than one signalling link in each link set, and/or have a redundant signalling route. During dimensioning, each of the links should be over-dimensioned, so that if there is a failure, the link should be able to carry extra traffic. Thus, there has to be a trade-off between the resources and failure protection in core networks.

Output of Network Dimensioning

As noted above, network planning tools are used for dimensioning and planning of a core network. Some results of the dimensioning exercise are explained below

- Traffic calculations: Based on inputs such as the number of subscribers, calling interest, moving interest, forecast number of subscribers, etc., a traffic matrix is generated.

- Number of switches, their capacities and locations: The number of switches required is one of the main outputs of dimensioning. This is based upon subscriber capacity, geographical conditions, types of equipment available (in terms capacity), the type of routing employed, etc.

- Transmission connections: This refers to the interconnections between the switches and the actual capacities required for these transmission links. Transmission can be via optical fibre cables or microwave links (for dimensioning and designing the latter, refer to Chapter 2).

- Signalling plans: The various aspects of signalling such as routing, protection and related synchronisation should be planned during the network analysis and dimensioning phase.

These plans constitute the major outputs of the dimensioning phase. Some typical examples of switch and signalling network dimensioning plans are shown in Figures 4.8 and 4.9 respectively.

Figure 4.8 shows an example of a big network. Instead of interconnecting the switches directly, transit switches have been used. Each region generates traffic from mobile, fixed

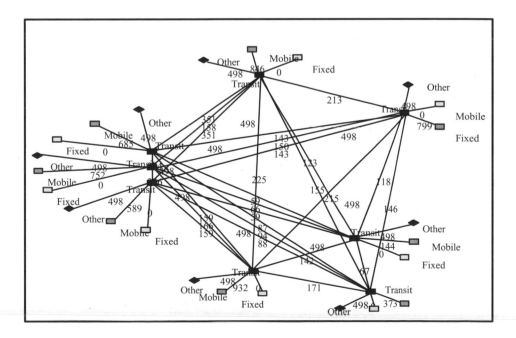

Figure 4.8 Typical output of switch network dimensioning

and other sources. The traffic is routed through the transit switches. The numbers shown in the figure indicate the traffic in milli-erlangs (mErl).

Figure 4.9 shows a typical plan with signalling links and link sets (SLS), along with routing plans. This is a simplified example; in practice, more complicated networks involve more switches, and signalling plans become more complicated.

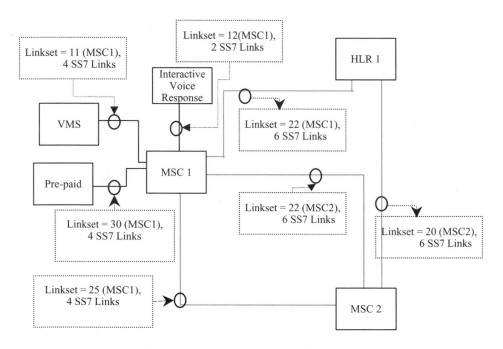

Figure 4.9 Example of a signalling plan

4.6 DETAILED PLANNING

Detailed planning of a core network generally consists of signalling plans, routing plans, numbering and charging plans, DCN settings, synchronisation plans, etc. This information arises from the network analysis as already described. The major outputs of the detailed core network plan are described below.

- Routing plans: Types of routing are decided in the analysis and dimensioning phase. Now, routing is fixed, names and naming conventions are adopted and created for destinations and sub-destinations, circuit groups, etc. This is explained in detail in section 4.8.

- Signalling plans: Signalling points, SEP, STP, etc., are defined. Signalling link numbers and signalling link sets are also finalised. Signalling routes are defined, keeping in mind the protection aspects.

- Numbering and charging: The numbering group used by each switch is finalised. This usually takes into account geographical locations in order to make it easier. Charging zones are defined, along with methods of collection and record transfers. The main categories of numbering plans are IMSI (International Mobile Subscriber Identity), MSISDN (Mobile Subscriber ISDN number), value-added services number (e.g. in a virtual private network), roaming numbers, handover numbers, test numbers, emergency numbers, etc.

- Synchronisation plans: Synchronisation is defined at switch level along with the priority settings. The principles of synchronisation were defined in Chapter 2.

The source data for each of these aspects is generated for each switch. An example of a detailed plan for SLS is shown in Figure 4.10.

Signalling Link Sets {SLS}											
SIGNALLING NETWORK TYPE	SIGNALLING LINK SET NUMBER	SIGNALLING LINK SET NAME	SIGNALLING POINT CODE	SIGNALLING POINT NAME	SIGNALLING LINK NUMBER	SIGNALLING LINK CODE	SIGNALLING LINK PRIORITY	EXTERNAL PCM - TSL	CONTROLLING UNIT		REMARKS
NA0,NA1	NUMERIC	IDENTIFIER	HEX / DEC	IDENTIFIER	NUMERIC	NUMERIC	NUMERIC	NUM-NUM	TYPE	NUMBER	INFO
IN0,IN1	0...299	5 CHARACT.	0..3FFF/0....16383	5 CHARACT.	0...254	0...15	0...15	64...511 - 1...31	CCSU	0...	
NA1	0		100H	DUS01	0	0	0	130-16	CCSU	0	
NA1	1		150H	FRA01	1	0	0	132-16	CCSU	1	

Figure 4.10 Detailed plan for SLS

4.7 CORE NETWORK OPTIMISATION

4.7.1 Basics of the Optimisation process

Core network optimisation focuses on both switch and signalling network optimisation. The process starts with the defining of key performance indicators (KPIs) along with data collection. This is then followed by analysis of the data. A performance analysis is done for both the switches and signalling. In some cases, the transmission network (for the core) and other aspects such as IN may be included and their effects on the core network might also be included. Analysis of the network and data might bring to light some bottlenecks, and lead to suggestions for a network upgrade, or parameter tuning. These corrections are then made/suggested in the final core network optimization plan for implementation (see Figure 4.11).

Figure 4.11 Core network optimisation process

Any information that is relevant to the quality of service of the network can be considered to be part of the key performance indicators. This may include parameters related to traffic and signalling performance, and measurements related to HLR, VLR and other network elements in the core network. Data can be collected by using test equipments, by using counters on the NMS, or by means of other measurements/reports from the network.

4.7.2 Data Collection and Analysis

Data Collection: Switching

In Chapters 2 and 3, we have seen that data on the existing network topology is required. Similarly in core network optimisation, data on the existing network topology should be collected and analysed. This mainly consists of information about existing network elements in the core network and interconnections between them (as this will be the basis of routing analysis and optimisation).

Optimisation in the core network is dependent mainly on traffic measurements, which is also the most critical input for the whole process. The data generally contains information such as cell measurements, traffic category measurements, measurements related to the busy hour, incoming traffic, outgoing traffic, etc. Knowledge in these areas aids an understanding of the condition of the system, and helps with further recommendations to alleviate problems with existing and anticipated traffic. The network in general should have nomenclature consistency. The inputs for this generally consist of the naming of network elements (in the core network), the naming of destinations and sub-destinations, the naming of routes, etc. Another input required is configuration parameters of the core network elements, mainly the MSC/VLR and HLR. From the MSC/VLR, this consists of parameters related to charging, roaming, authentication, ciphering, etc., while from the HLR, parameters related to transferring of subscriber data, basic and supplementary services, etc., are required.

Unlike in radio and transmission network optimization, where tools other than the NMS are also used for data collection, in the core network optimisation process the NMS plays a more important role. However, the type of measurements that need to be done should be created in the switching platform. These measurements include the network elements and/or part of the system that is connected to a switch. A typical example is a control group measurement that covers all the control units in the core network, as shown in Figure 4.12.

The schedule and duration of the measurements need to be specified. Planning engineers should take into consideration the system's capability and requirements for such measurements. Every equipment manufacturer has its own specification such as the number

SWITCHING PLATFORM LONDON			29		- 0- 1999 02:02:29			
TRAFFIC MEASUREMENT REPORT								
SOURCE: STU - 1 OUTPUT INTERVAL: 45 MIN								
UNIT	CALLS	ACCEP	ANSW	SFAIL	IFAIL		EFAIL	ERLANGS
LSU-0	0	0	0	0	0		0	0.0
LSU-1	0	0	0	0	0		0	0.0
CCSU-0	0	0	0	0	0		0	0.0
CCSU-1	121	121	121	0	0		0	1.4
CCSU-2	633	633	633	0	0		0	12.6
BSU-0	0	0	0	0	0		0	0.0
BSU-1	1890	1890	1890	0	0		0	21.4
BSU-2	3139	3139	3139	0	0		0	26.9
IWCU-0	0	0	0	0	0		0	0.0
IWCU-1	0	0	0	0	0		0	0.0
END OF REPORT								

Figure 4.12 Example of control group measurements

of measurements available, the number of measurements that can run at a given time, limitations on the duration of measurements, and delivered outputs (in ASCII/Word/text files, etc.).

Data Collection: Signalling

The existing signalling plan is the basis for the signalling optimisation process. Information about signalling links, signalling link sets, capacity of the signalling links, signalling routes, signalling route sets, signalling network topology, signalling load and signalling load sharing is important for this process. Some of the information, e.g. signalling topology, is available already, but for some – such as the existing load – measurements need to be performed. Again, the NMS can be used to collect information related to the performance of the signalling links. This can be done with the help of statistical counters available for monitoring performance of the signalling network. One such example of signalling link utilisation is shown in Figure 4.13.

If possible, measurements and statistics from all the signalling elements in the core network should be fetched. These measurements and statistics are analysed and compared with the existing plans, and changes are suggested if needed.

```
METERS OF LAST PERIOD: 01:45:00-02:15:00 (30 MIN)

         3.1         3.2        3.3         3.4         3.5
  LINK   3.6    L1   L2         L3    TOT
  &      3.7    L1   L2         L3                TOT
  BIT    3.10   L1   L2         L3                TOT
  RATE   3.11   L1   L2         L3                TOT
  =====  ========== ========== ========== ========== ==========
    1    0000008071 0000000000 0000000170 0000014343 0000000173
   64K   0000000000 0000000000 0000000000 0000000000
         0000000000 0000000000 0000000000 0000000000
         0000000000 0000000000 0000000000 0000000000
         0000000000 0000000000 0000000000 0000000000
```

Figure 4.13 Statistics showing signalling link utilisation

Data Analysis: Switching and Signalling

The data collected now needs to be analysed so that suggestions for improvements and optimisation can be given. Traffic measurement reports are the inputs for the analysis. Usually, traffic generated is divided into categories, as seen in the planning phase. The categories include traffic originating and terminating in the network, traffic originating and terminating in an external network, internal and transit traffic, etc. This categorisation makes the analysis much easier. The traffic and signalling analysis will result in information such as:

- traffic handled by the switches/exchanges

- the exact amount of traffic under each traffic class

- subscriber calling-related measurements (subscribers/calls/successful call attempts/ traffic intensity)

- traffic loading in the switching exchanges and their availability (leading to congestion figures)

- configuration of the signalling network

- loading on the signalling network.

The analysis is usually done using network planning tools (refer to Appendix A). Suggestions for improvements are made based on the analysis outcome.

4.7.3 Core Network Optimisation Plan

Switching Optimisation Plan

- If a congestion problem is identified, extra PCM connections should be suggested at the location where it is experienced. If the congestion is severe and a new switch is required, then it should be proposed. A whole network topology should be produced, with information on the location of the new switch, traffic routed through it, etc.

- Inter-switch connections, and traffic routing between the MSC and transit switches, should be modified in locations/regions where a transit switch is carrying excessive traffic. This would mean devising new routing plans.

- When networks are rolled out, mismatch in naming conventions may happen. One of the objects of the optimisation process is to clear the 'naming mess'. Naming conventions should be applied in a manner that it uniform in the whole network.

Signalling Optimisation Plan

- The number of signalling links should be optimised, with increases or decreases as required.

- The signalling links and sets should be distributed uniformly across the network. If this is not the case, new signalling links and link sets should be proposed.

- Usually load sharing is not equal. New signalling plans and network topology should be created to remedy this.

- Proposals should be made for redundancy in the signalling control units.

II

2.5-generation Network Planning and Optimisation (GPRS and EDGE)

5

GPRS: Network Planning and Optimisation

5.1 INTRODUCTION

GSM was capable of providing a data rate of 9.6 kbps on a single time slot. With the advent of high-speed circuit-switched data (HSCSD), the capability of the network was increased multi-fold, to 115.2 kbps. In practice, however, it was only 64 kbps owing to the limitation of the A-interface and the core network. The main benefit of the implementation of HSCSD was that, with limited upgrading (i.e. minimum investment), the capacity for data transfer was increased to up to four TS on the receiving side and two TS on the transmitting side. But the traffic was still circuit-switched, which meant a long access time to the network. As charging is proportional to the logging time, the subscriber ends up paying more. This led to the evolution to the packet-switched network. In this technology, the access time to the network is reduced and charging is done solely on the usage of the network; i.e. even when a connection is there but not being used, the subscriber is not charged. Usage of the network resources becomes more dynamic and efficient. They are no longer reserved for a user logged to the network, even when he is not using the resources. This system was known as a *general radio packet system* (GPRS).

GPRS is an addition to the existing GSM system, enabling packet-switched transmission in the network whilst keeping the existing value-added services like SMS, etc. Because of this, data rates increase substantially: the user now can log into the GPRS network, and can make use of all eight TS dynamically and be charged only when using the resources. The packet data can be sent during idle times also, between speech calls, thus making effective use of the network resources and saving money for the subscriber.

Fundamentals of Cellular Network Planning & Optimisation A.R. Mishra.
© 2004 John Wiley & Sons, Ltd. ISBN: 0-470-86267-X

5.2 THE GPRS SYSTEM

GPRS technology is an addition to the existing GSM technology. Because of the introduction of packet switching, the new network elements are those capable of performing packet switching. The main ones are the *serving GPRS support node* (SGSN) and the *gateway GPRS support node* (GGSN).

The GSM system is orientated towards providing a voice service. So, apart from the addition of new elements such as SGSN and GGSN, there are only minor changes required in the GSM network elements in the BSS and HLR. These are both hardware- and software-related changes and are due to the higher-level coding schemes that are being used in the GPRS technology. The most important change is the addition of a PCU (packet control unit) at the base station controller. The GPRS system with all these elements looks the same as the GSM except for the addition of the packet-handling core part, as shown in Figure 5.1.

As all the network elements of the GSM have been explained in earlier chapters, here only the new elements will be discussed. But first let us look at the changes in the mobile station.

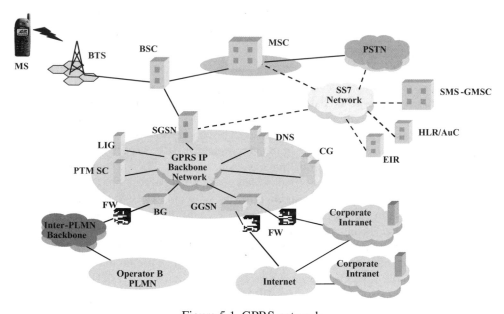

Figure 5.1 GPRS network

GPRS Mobile Station

The fundamental difference between a GSM mobile and GPRS mobile (shown in Figure 5.2) is that the GPRS mobile is able to handle the packet data at a higher speed.

GPRS mobile stations have been classified into three classes, A, B and C, based on their ability to handle cellular networks. Class A mobiles are connected to both the GSM and GPRS networks and can use them simultaneously. Class B mobiles are connected to both the networks, but they can use only one at a time. Class C mobiles can be connected to either one of the networks.

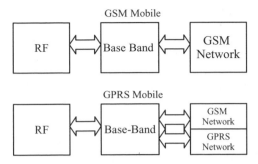

Figure 5.2 GSM and GPRS mobile station

Serving GPRS Support Node (SGSN)

SGSN is the most important element in a GPRS network. It is the service access point for the mobile station. Its main functions include mobility management and registration and authentication. It also interacts with a mobile with packet data flow and functions related to it like compression and ciphering. These are handled by protocols such as the SNDCP (sub-network dependent convergence protocol) and LLC (logical link control). SGSN is also responsible for GTP (gate tunelling protocol) tunnelling to the other support nodes.

Gateway GPRS Support Node (GGSN)

The GGSN is connected to the SGSN on the network side and to the outside world external networks such as the Internet and X.25. As it is a gateway to the external networks, its main function is to act as a 'wall' for these external networks in order to protect the GPRS network. When data come from the external network, after verification of the address, the data are forwarded to the SGSN. If the address is found to be invalid, the data are discarded. On the other hand, the SGSN also routes the packets it receives from the mobile to the correct network. Thus, for the outside networks, the SGSN acts as a router.

Border Gateway (BG)

The border gateway interconnects different GPRS operators' backbones, thereby facilitating the roaming feature. It is based on the standard IP router technology.

Legal Interception Gateway (LIG)

The LIG performs 'legal' functions in the network. Subscriber data and signalling can be intercepted by using this gateway, thus enabling the authorities to track criminal activities. LIG is required when launching a GPRS service.

Domain Name System (DNS)

DNS does the translation of IP host names to IP addresses, thereby making IP network configuration easier. In the GPRS backbone, SGSN uses DNS to get GGSN and SGSN IP addresses.

Packet Control Unit (PCU)

This is a new card that is implanted in the BSC to manage the GPRS traffic. The PCU has limitations in terms of the number of transceivers and base stations it can manage, thereby creating a bottleneck for the network design usually in terms of capacity. Increasing the capacity of the network leads to an increase in PCU capacity, thereby increasing the hardware costs of the network.

5.3 INTERFACES IN A GPRS NETWORK

Owing to the addition of extra network elements, some new interfaces are added in the GPRS network. All these new interfaces are known as G-interfaces, as shown in Figure 5.3. A brief description of these interfaces is given below.

- G_b interface: This lies between BSS and SGSN. It carries the traffic and signalling information between the BSS (of GSM) and the GPRS network, thus easily making it the most important interface in network planning.

- G_n interface: This is present between the SGSN and SGSN/GGSN of the same network. It provides data and signalling for intra-system functioning.

- G_d interface: This is present between the SMS-GSMC/SMS-IWMSC and SGSN, providing for better use of the SMS services.

- G_p interface: This lies between the SGSN and the GGSN of other public land mobile networks. Therefore it is an interface between the two GPRS networks. This interface is highly important considering its strategic location and functions that include security, routing etc.

- G_s interface: This is present between the SGSN and MSC/VLR. Location data handling and paging requests through the MSC are handled via this interface.

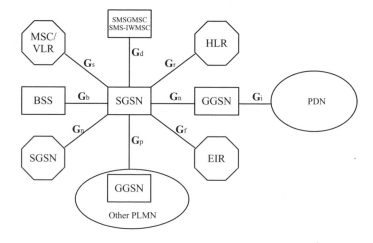

Figure 5.3 GPRS interfaces

- G_r interface: As this is an interface between the SGSN and HLR, all the subscriber information can be assessed by the SGSN from the HLR.

- G_f interface: This interface gives the SGSN the equipment information that is present in the EIR.

- G_i interface: This lies between the GGSN and external networks. This is not a standard interface, as the specification will depend on the type of interface that will be connected to the GPRS network.

5.4 PROTOCOL STRUCTURE IN A GPRS NETWORK

The protocol structure in GPRS is quite different from that for GSM, as shown in Figure 5.4.

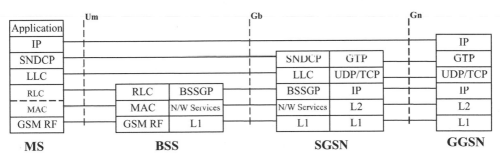

Figure 5.4 GPRS protocol stack

5.4.1 MS Protocols

Two kinds of physical layer have been defined in the GSM 05 specifications: the *RF layer* and the *link layer*. The physical RF layer performs modulation/demodulation of the input signal (information) apart from FEC (forward error correction) coding, interleaving and congestion detection. The physical link layer provides the services that are required to transfer this information over the air interface; these include coding, framing, synchronisation, monitoring of radio quality, power control procedures, and transmission error detection and correction. The physical link layer can support multiple mobile stations that share the same physical channel.

- The RLC/MAC layer is present over the physical RF layer. The RLC (radio link control) is responsible for data transmission over the physical layer of the GPRS radio interface. Also, it provides a link to the upper layers. The MAC (medium-access control), as the name suggests, is responsible for control of the access functions of the mobile over the air interface. These functions include the allocation of channels and multiplexing of resources.

- The LLC (logical link control) layer is present over the RLC layer and, as the name suggests, it is responsible for the creation of a logical link between the mobile and the SGSN. Thus, is responsible for the transfer of the signalling and data transfer. It is independent of the lower layers.

- The SNDCP (sub-network dependent convergence protocol) layer performs mapping and compression functionalities between the network layers and lower layers. Compression of both the control and data information is done by this layer. It also performs the segmentation/de-segmentation of the information to/from the lower LLC layer.

The above-mentioned are responsible for the information flow between the MS and SGSN. The IP layer is responsible for the formation of the backbone of the GPRS network and a direct interface between the MS and GGSN.

5.4.2 BSS Protocols

The BSSGP (base station subsystem GPRS protocol) layer is responsible for information transfer between the SGSN and the RLC/MAC layer. Thus, its primary function is to create the environment for data flow between these two entities.

5.4.3 SGSN protocols

The GTP (GPRS tunnelling protocol) layer runs over the UDP layer and is responsible for the data and signalling information between the BSC and the GSN nodes. This protocol forms a tunnel for each subscriber and each tunnel is identified by a tunnel endpoint identifier.

The UDP/TCP (user data protocol/transport control protocol) is the backbone network protocol that is used for routing network data and control signalling. While the TCP is a connection-oriented protocol providing reliable data transmission service, UDP is a connectionless protocol providing data transmission services that are unreliable.

The IP (Internet protocol) is the datagram-oriented protocol. Both UDP and TCP interface with the IP directly. The major function of the IP is as a routing protocol that provides a means for devices to discover the topology of the network as well as detect charges of state in nodes, lines and hosts.

5.5 GPRS NETWORK PLANNING

As already stated, the main difference between GSM and GPRS networks is the addition of a packet-data handling capability in GPRS. All the differences in network planning are due to this additional aspect. Radio network planning and core network planning are the aspects most affected, while transmission network planning experiences minimal changes.

5.5.1 Radio Network Planning

The radio network remains largely the same as with GSM (as described in Chapter 2). Some of the aspects change owing to the introduction of packet data. This mainly affects the dimensioning and detailed planning, which directly affect the coverage and capacity planning, leading to an impact on radio network quality. In GSM, radio network quality meant voice quality, but in a GPRS network quality includes both voice and data quality. This leads to changes in the key performance indicators (KPIs). In the following, only the areas that need changing with respect to radio planning in GSM are explained. Fundamental concepts include the following:

- logical channels
- coding schemes
- management: RRM and MM (radio resource and mobility management)
- resource allocation
- power control.

Logical Channels

Owing to the involvement of data packets, some new channels are used. A separate set of channels is allocated for packet data, allowing more flexibility in the necessary signalling. These channels (see Table 5.1) are mapped into the physical packet data channel (PDCH). Different logical channels can find their place on one single physical channel.

Table 5.1   GPRS Logical Channels

Channel	Abbreviation	Function/Application
Packet Broadcast Control Channel (*DL*)	PBCCH	Broadcast system information specific to packet data
Packet Common Control Channel	PCCCH	Contain logical channels for common control signalling
Packet Data Traffic Channel	PDTCH	Channel temporrily used for data transfer.
Packet Associated Control Channel	PACCH	Used for signalling information transfer for a given mobile
Packet Access Grant Channel (*DL*)	PAGCH	Notifies that mobile about resource assignment before actual packet transfer
Packet Notification Channel (*DL*)	PNCH	Used for sending information to multiple mobile statios.
Packet Paging Channel (*DL*)	PPCH	Pages a mobile station before packet transfer process begins
Packet Random Access Channel (*UL*)	PRACH	Used by the mobile station for initialisation of the uplink packet transfer.

Coding Schemes

The radio block consists of header, data and control information, which are basically the MAC header, the RLC data block and MAC/RLC control information. RLC data are encoded for security. In GPRS systems, there are four coding schemes that are used for the packet data: CS-1, CS-2, CS-3 and CS-4. Table 5.2 shows the coding schemes and related parameters.

Table 5.2   Coding Scheme in GPRS

Coding Schemes	Code Rate	Data Rates (kbps)	Data Rates (kbps) (excl. headers: RLC/MAC)
CS-1	1/2	9.05	8
CS-2	$\sim 2/3$	13.4	12
CS-3	$\sim 3/4$	15.6	14.4
Cs-4	1	21.4	20

In coding scheme CS-1, half-rate convolution code is used for forward error correction (FEC). It has a data rate of 9.05 kbps. Coding schemes CS-2 and CS-3 are the same as CS-1 but in punctured format. Puncturing is done so as to increase the data rate but it comes at the cost of reduction in redundancy. Coding scheme CS-4 has 'no coding' (i.e. no FEC), so further increasing the data rate to 21.4 kbps.

Radio Resource and Mobility Management

In the GSM system, IDLE and DEDICATED are the two states of a mobile station. In a GPRS system there are three states: IDLE, STANDBY and READY. In the IDLE mode, a subscriber is not attached to the GPRS network, while in the STANDBY mode the GPRS network knows the routing area location of the mobile station. The mobile station enters the READY state by sending a service request to the network. In this state, the mobile station becomes attached to the GPRS mobility management and is known by the network on the call basis. Once the call is finished, the mobile enters into the STANDBY mode again. Thus, when moving from STANDY to READY, the SGSN receives and processes a *GPRS attach request*; and when moving from READY to STANDBY, the SGSN receives and processes a *GPRS detach request*.

For movement of data between the MS and the GPRS network, the PDP (packet data protocol) context needs to be activated. Either the MS or the GPRS network can generate the PDP context. These requests are sent and processed by the SGSN. A PDP context generally contains activated information such as addresses (required for the data traffic), QoS-related information, protocol types, etc. MS uses the PDP contexts in the STANDBY and READY states. The number of PDP contexts is one of the factors that have a direct impact on the SGSN capacity.

Resource Allocation

Radio time-slot allocation also changes in the GPRS air interface. It becomes more dynamic. A GPRS mobile is capable of using the network both for voice (CS) and data (CS and PS). GPRS traffic is managed by the BSC as it does the allocation of resources for the CS and PS data. The time slots that handle the CS traffic fall under *CS territory* and the time slots that handle PS traffic fall under the *GPRS territory*. Some of the time slots are in dedicated mode and some in default. Each group of time slots is known as a *territory* (see Figure 5.5).

Consider a base station with two transceivers, TRX1 and TRX2. There are eight radio time slots in each of the two TRXs. In TRX1, two TS are allocated to the signalling (i.e.

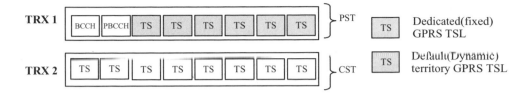

Figure 5.5 GPRS timeslot allocation to CS and PS traffic

BCCH and PBCCH). Of the remaining six, three fall under dedicated territory and three under default territory. The TS that are to be used only for the packet data (so cannot be used for the CS traffic) are known as dedicated time slots and fall under dedicated territory. The remaining three TS can be used for both voice and data and fall under default territory.

The eight time slots that are used for voice traffic come under the CS territory. Consider a case when all the eight TS in TRX2 are occupied and there is a ninth call; then one TS from the default territory is assigned to this CS traffic. Once the traffic in TRX2 decreases, the call on the default will be switched back to CS territory, leaving the default time slots for possible PS traffic.

Power Control

Power control in GPRS networks is more complicated because of the addition of the PS traffic. Power control takes place in both the uplink and downlink directions. The uplink power control is used to reduce interference and helps in increasing the battery life of the mobile station. In the downlink direction, the power control feature is used to control the output power of the base station and subsequently help in decreasing the interference in the network. As in a GSM system, the mobile station performs the measurements and, based on these, the output power of the base station is controlled.

Temporary Block Flow (TBF)

This is a new concept introduced into GPRS networks. The network and mobile establish a connection for data flow. This connection is unidirectional in nature and is maintained for the duration of the call. It is established for the packet data (i.e. blocks) and is not permanent (hence temporary block flow). It can be uplink, downlink or simultaneous uplink and downlink.

5.5.2 The Radio Network Planning Process

Both GSM and GPRS have the same radio-wave propagation principles. The one important extra consideration is the addition of data traffic. Pre-planning and nominal planning steps, such as site survey and site selection, remain the same as with GSM, as explained in Chapter 2. Detailed planning of the GPRS network takes account of the data traffic and the new equipment added to the network for this. The detailed planning again focuses on coverage, capacity, frequency and parameter planning.

Coverage Planning

Coverage in a GPRS network depends are the S/N (signal-to-noise ratio) and data transmission rates. Interference can be a limiting factor for the maximum data rate in the network. Each of the coding schemes (CS-1, CS-2, etc.) works for a certain range of C/I (channel to interference ratio) for a given value of block error rate (BLER).

The coverage plans are made with the objective of providing a balanced link budget for both the uplink and downlink directions. The link budget is similar to that of a GSM radio network. However, the threshold requirements change in GPRS networks, thereby giving a different coverag area by a cell. The two main parameters that are required for link budget calculation are the transmitted power (from mobile station and base station) and the receiver sensitivity. As the coding schemes in a GPRS radio network have different S/N requirements, the areas covered will be different. Coding scheme CS-1 covers a large area compared to scheme CS-4. One important change that is seen in the link budget calculations is the removal of a body loss value for CS-2, thereby giving GPRS services a 3 dB advantage compared with GSM services. The S/N requirements for CS-3 and CS-4 are quite high, thereby reducing the area covered by them. CS-1 and CS-2 are usually used for GPRS radio network coverage planning, while CS-3 and CS-4 are used for the call centre. As coverage in a GPRS network is limited by interference (rather than by noise), the C/I ratio distribution becomes a detrimental factor in coverage area predictions.

Capacity Planning

Capacity planning of a GPRS network may be subdivided into two parts: capacity planning for the radio interface and capacity planning for the G_b interface. In this section, we deal with capacity planning for the radio interface.

The network has three kinds of traffic: voice, CS data and PS data. All these have to be considered when doing capacity planning for the radio interface. Circuit-switched traffic always has priority over PS traffic, but owing to the delay-sensitive nature of some PS services, some time slots are *dedicated* to carry the PS traffic only.

CS traffic calculations, as with GSM, pre-dominantly involves the Erlang B tables, blocking and C/I thresholds. Assume the case shown earlier in Figure 5.5. There is one cell that has two TRXs. In ideal conditions (i.e. without blocking), 14 (voice) users can use the time slots continuously, so traffic of 14 Erl would be generated if there is no blocking. If the number of voice users is reduced to eight, then the remaining six time slots can be used for data. It should be noted, however, that data which are not delay-sensitive could still be sent through the *gaps* in the air interface. Only the data that are delay sensitive need uninterrupted availability of time slots.

When an existing GSM network is upgraded to a GPRS network, the available capacity falls short for the PS data. Increasing the number of TRXs and the time slots in GPRS territories (dedicated + default) would be one effective way to tackle the capacity problem.

Quality of service (QoS) has a deep impact on capacity planning. An increased load would decrease the quality of a call. For critical applications, a minimum QoS should be met, which means that the loading can increase only up to a certain point. Thus, frequency planning takes an important place in achieving a desired QoS level in a GPRS radio network.

Frequency Planning

Coverage and capacity planning go hand-in-hand, and coverage planning is quite related to frequency planning. An effective frequency plan will increase coverage areas significantly and limit interference (as will power control).

The principles and methods of frequency re-use (the same as used in GSM radio networks, e.g. frequency hopping) are used extensively so that the spectrum is used effectively. Power control is more necessary in the downlink direction. Using the BCCH layer does this. The BCCH layer has an important characteristic, namely that burst transmission in the DL is constant and with full power. This means that variation in throughput is due to user multiplexing over the same time slot, thereby making the time slot capacity constant (and independent of the GPRS traffic load). Time slot capacity is also interference-limited. With an increase of traffic, the number of users per time slot decreases because of interference.

Thus, in a GPRS network, interference reduction becomes quite an important aspect of the whole network-planning scenario.

Parameter Planning

Parameter planning in a GPRS network can be considered to be an extension of GSM parameter planning. Signalling, RRM, power control, handover, etc., are still relevant, and extra parameters related to packet data are added.

The major enhancements are in the signalling parameters. As seen earlier in Table 5.1, there is whole list of parameters associated with the signalling of packet data transfer. One important parameter to decide is whether or not the GPRS traffic goes on the BCCH time slot. Then there are parameters that are related to defining the GPRS territory. Lastly, there are parameters related to routing and location area codes, to ensure enough capacity is available for paging.

5.5.3 Transmission Network Planning

The fundamental concepts remain the same as for a GSM network, as described in Chapter 3. However, one extra aspect that transmission planning engineers have to deal with is PCU dimensioning. The packet control unit is located in the BSC, as shown in Figure 5.6. It is responsible for management of the GPRS (or packet) traffic.

The main aspect of dimensioning of the PCU is its capacity with respect to:

- the maximum number of PDP contexts

- the maximum number of TRXs

- the maximum number of BTSs

- the maximum number of PCM lines towards the A_{bis} and towards the G_b interface, as well as the traffic on the G_b interface

- the maximum number of location areas and routing areas.

BSC

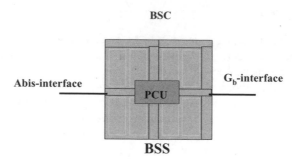

Abis-interface G_b-interface

PCU

BSS

Figure 5.6 Packet control unit (PCU)

5.5.4 Packet Core Network Planning

Owing to the introduction of packet data, core network planning is subdivided into two major parts: planning for the circuit core and for the packet core. Planning for the circuit core remains more or less as discussed in Chapter 3, so will not be repeated here. This section covers planning for the packet core.

The most important aspect of packet core network planning is dimensioning of the three interfaces, G_b, G_n and G_i (refer back to Figure 5.3).

The G_b interface

The interface between the BSS and SGSN, which allows the exchange of both data and signalling information, is called the G_b interface, as shown in Figure 5.7. This interface is more dynamic than the A-interface. It not only allows multiple users to share its resources, but also reallocates the resources once the data transfer is stopped, as compared to the A-interface where the physical resources are dedicated to the user irrespective of their usage.

The protocol stack of the G_b interface is shown in Figure 5.8. There are three layers on the top of the physical layer. The physical layer serves the upper layers and transfers the data and signalling information from one end to another along with the overheads that are generated by each of the layers. The interconnection between the BSS and the SGSN can be done by using any of the physical media or by implementing the frame relay network. The advantage of using the frame relay network is that the interfaces at the two ends can be different.

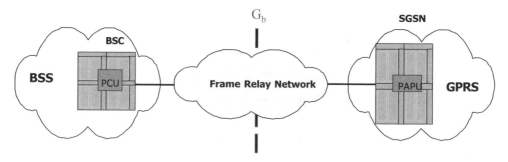

Figure 5.7 G_b interface

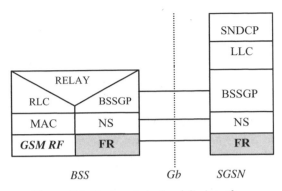

Figure 5.8 Protocol stack of G_b interface

A frame-relay network is basically a packet-switching network that transfers the (packet) data in packets of variable lengths. These packets are known as frames. The network service (NS) layer controls the frame-relay layer.

There are two hardware units associated with the G_b interface or the frame-relay network: PCU and PAPU. The packet control unit (PCU) is located on the BSS side, while the packet processing unit (PAPU) is located on the SGSN side. For G_b interface dimensioning there are three major aspects: BSC (or PCU) dimensioning, SGSN dimensioning, and frame relay planning. We have already seen the inputs that are required for PCU dimensioning. For BSC dimensioning from the perspective of packet core network planning, the only input required is the BSC's PCU handling capacity.

SGSN Dimensioning

SGSN dimensioning basically results in the number of SGSNs that may be required in a GPRS network. Major inputs are the SGSNs own capacity in terms of handling the amount of data, subscribers and the number of PCMs.

The data handling or the processing capacity will determine the number of SGSNs required in a given network based on the amount of traffic. Calculations should not be done based on the traffic generated by the total number of BSCs, because at a given moment all the BSCs will not be fully loaded. Similarly, not all subscribers will be using the GPRS at the same time. So, a reduction factor (i.e. what percentage of users will be using the GPRS network at the same time) should be planned carefully. Apart from this, every SGSN (depending on the vendor) has its own capacity for handling the number of frame-relay links.

Frame-relay Planning

There are many ways in which to transfer data from the BSC to the SGSN. This may include techniques such as sending voice and data traffic separately, or multiplexing both the voice and data traffic. In the initial phases of network launch, the PS data might be negligible in terms of the amount of data handled by each PCU. Thus, the traffic from some PCUs is then multiplexed and transmitted to the SGSN. The method of transmission can be similar to that from BTS to BSC, or BSC to MSC, or it can be sent by the frame-relay method. If there is no frame-relay network, then the ATM network could also be used to transfer the frame relay. Even dedicated PCM links (also known as frame relay links) could be used to send packet data from the BSC to the SGSN. The main result of frame-relay dimensioning is to find the number of time slots that are needed to send the packet data traffic from the BSC to the SGSN.

The G_n interface

The interface connecting all the GPRS network elements (SGSN–SGSN, SGSN–GGSN, etc.) is the G_n interface. Dimensioning of this interface is based on the amount of data flow during peak hours, the number of subscribers, the number of G_b interfaces, information about the IP network, etc. The core planning engineers classify the GPRS network based on this information.

Generally the GPRS network could be one of a number of types. It could be a GSM network in which the GPRS network elements are introduced to launch the packet data services, or it could be an advanced network that focuses on issues like coverage and quality. The main difference between these two types of GPRS network is in the GPRS network elements that would be used. In an advanced network, the number of SGSNs or GGSNs would be increased, thereby increasing the number of G_b or G_n interfaces. An Ethernet LAN connection is the GPRS IP backbone. All the network elements inside the GPRS packet core network – such as NMS, firewalls, DNS, SGSN, GGSN, etc. – are connected through this switch. Thus, it becomes necessary to not only use a good-quality switch but also keep a redundant switch.

The G_i interface

The core network planning engineers have also to look into the interface between the GPRS network and any external network (i.e. the interface between the GGSN and the external network). Since a different type of data will be flowing through this network, G_i is not a standard interface but acts as a reference point.

As the G_i interface interacts with external networks, security is also an important issue. Firewalls are used between the GGSNs and external networks. A firewall protects both the subscribers and the QoS of the network (by not letting it become overloaded).

Connection between the GGSN and an external network is done through virtual routers (also known as access points). A gateway tunnelling protocol (GTP) in the GPRS network is responsible for routing of the information either way.

Another important aspect that is taken care of by the core planning engineers is IP addressing of the network elements. Some fundamentals of IP addressing and routing are given below.

Basics of IP addressing and Routing

IP Addressing
Data exchange between two network elements is a three-step process: addressing, routing and multiplexing. Every network element in the Internet should have a unique IP address, making it possible for data to reach the right host. Routers target the data to the correct network, while multiplexing makes it possible for the data to reach the correct software within the host.

There are three classes of IP address, A, B and C. The number of networks in each class can be computed as:

- class A: 128

- class B: $64 \times 256 = 16\,128$

- class C: $32 \times (256)^2 = 2\,097\,152$.

The number of addresses in each class can computed as:

- class A: $(256)^3 = 16\,777\,216$
- class B: $(256)^2 = 65\,536$
- class C: 256.

Thus, class A has the least number of networks and the largest number of addresses, while class C is has the largest number of networks and the least number of addresses. How can one identify to which class an address belongs? The addresses are of 32-bit length. If the address starts with bit 0, then it is a class A address. Class B addresses start with bits 1,0; while class C addresses start with 1,1,0.

- Class A: The first bit 0 acts as class identifier; the next seven bits as network identifier; the final 24 bits as host identifier.

- Class B: The first two bits 0,1 act as class identifier; the next 14 bits as network identifier; the final 16 bits as host identifier.

- Class C: The first three bits 1,1,0 act as class identifier; the next 21 bits as network identifier; the final eight bits as host identifier.

The IP addresses are written as four decimal numbers separated by points, with each of the four numbers ranging from 0 to 255. If the value of the first byte is less than 128, it represents class A addresses, 128–191 represent class B addresses, and 191–223 represent class C addresses. Addresses above the value of 223 are reserved addresses.

Another aspect to know of when doing IP planning is that of 'sub-network or subnet'. With the standard addressing scheme, a single administrator is responsible for managing host addresses for the entire network. As networks are growing rapidly, local changes may become impossible, thereby requiring a new IP address for the 'new' network. By using the concept of 'subnet-working', the administrator can delegate address assignment to smaller networks within the entire network. By sub-netting it is possible to divide the whole network into smaller networks in such a way that each of these smaller networks has its own unique address; and this address is still considered to be a standard IP address. The combination of the sub-net number and the host number is known as the local address. Subnet addressing is done in such a way that it is transparent to remote networks.

IP Routing
When the addresses have been defined, the next step is to route the traffic. Routing involves transportation of the packet data to the right host and via an optimal path. There are many parameters that determine the selection of the path, and there are algorithms that can be used to do the calculations. These algorithms generally need information such as the source address, the destination address and the next hop address. The core gateways have all the information about the networks that is necessary to find the optimal route for data. This is done through the gateway-to-gateway protocol (GGP).

For traffic routing in a complex network, a routing table is used. This contains all the information needed to route data to the required destination within the network, or to a local gateway if the destination address belongs to an external network.

5.6   NETWORK OPTIMISATION

Steps in the optimisation of a GPRS network are the same as for a GSM network (explained in Chapter 2) and involve definition of key performance indicators (KPIs), network performance monitoring and parameter tuning. However, GPRS network optimisation is more complicated because it uses the resources, such as frequencies, of the existing GSM network.

The optimisation process will focus on improvement in the following areas:

- Accessibility to the network: The mobile subscriber should be able to get the desired service when it is requested.

- Quality: A requested service should come along with the desired quality.

- Utilization of the resources: The air, A_{bis} and G_b interfaces are optimised.

- CS and PS: There needs to be improvement in both the CS and PS quality levels along with the improvement in coverage and capacity.

- Security: Security aspects of the network need to be reviewed.

The whole process can be divided into radio network, transmission network and core network optimisations. Only the radio and packet core network optimisation processes are covered here; see Chapters 3 and 4 for processes relating to the transmission and CS-core networks.

5.6.1   Radio Network Optimisation

As the GPRS (data) traffic runs on the GSM radio network, there are some complications that should be taken into account. Optimisation of the GPRS (data) network involves optimisation of the GSM radio (voice) network. This means there is a conflict between the priority of voice and data calls. Usually, voice calls take priority in such networks, so that the capacity/coverage and quality requirements of the mobile (data calls) subscribers are even more difficult to fulfil, leading to a continuous process of GPRS network optimisation.

Coverage in GSM and GPRS is dependent on C/I ratios. With the GPRS network using the resources of the GSM network, interference will be high, thus degrading the C/I ratio. Degradation of the C/I ratio in turn means a reduction in coverage areas. Moreover, addition of the GPRS network will decrease the voice quality of the network.

In urban areas, there is always a problem with frequencies. Channel allocation to a GPRS network in the initial phase is always based on the GSM channel allocation, with voice (CS) subscribers taking priority. Thus, capacity plans and channel allocations have to be studied and optimised again during this process.

Another aspect is the speed of the data services. Though the theoretical speed of a GPRS network should be about 171 kbps, in practice values are still around 40–60 kbps. The higher speed is possible only when all the eight time slots are utilised, which is not the case as usually five TS can be used for data. Moreover, the C/I requirements may not be satisfied for coding schemes beyond CS-1, thus reducing the data processing scheme.

All the above factors – coverage, capacity and quality – are also affected by the parameter settings. As seen in Chapter 2, parameter settings (or readjustments) again have an important role in the GPRS network optimisation process.

The optimisation process starts with the collection of required data. This can be radio network planning data from the design phase, and performance data.

Radio Planning Data

The existing network data includes site information such as the location of base stations, antenna locations and heights, antenna azimuths and inclinations (tilts), etc. Planning data also includes parameter settings. The configuration data includes base station numbers and identity codes (e.g. BSIC), cell configuration, frequency plans, power budget calculations (both uplink and downlink), etc. This information is similar to that required in GSM radio network optimisation. Additional information needed includes the radio time-slot usage for signalling, voice and data, as well as cell reselection algorithms, media access control, location areas, routing areas, handover criteria, number of subscribers (voice and data), processing and switching capacity, number of PDCHs supported simultaneously, etc. It would also be useful to have the data from any previous optimisation that was done for the GSM network.

Performance Data

Performance data collected usually relate to reliability and average throughput. From a user's perspective, reliability can be understood as success in call establishment and call release. The average throughput can be understood as the throughput with a desired quality as perceived by the subscribers.

Key Performance Indicators
Chapter 2 covered the voice KPIs of the radio network. Here we look at some of the data KPIs.

In a GPRS network there are a variety of services that can be present, with each service having a unique performance requirement. The quality of each service can usually be negotiated between the mobile station and the network, and should be maintained during the duration of the call. As mentioned above, reliability and average throughput are the two main criteria behind the process of defining the KPIs. Apart from these two, other important functions on which KPIs are based are load, utilisation, blocking and delay.

As reliability is about call establishment and release, the key performance indicators are blocking in the uplink and downlink directions both for PDTCH and TBF. For the average throughput, the uplink and downlink average throughput could be measured. For enhanced reliability, re-transmissions are done on the air interface (i.e. in the RLC layer). PDTCH channels carry the user data, re-transmission and control information. For loading, data Erlangs is used as a key performance indicator and is based on the number of radio blocks per duration of busy hour.

The amount of hardware used for data services can be indicated by the utilisation factor, which describes the average amount of GPRS time slots that are used for PS services. TBF blocking is another limiting factor in a GPRS network. As the value of TBF blocking goes higher, the network performance goes down. Delay is another important factor in data networks. Some PS applications are sensitive to delays, but as the loading of the network increases, the response time of the network increases, thereby increasing delays. These delays may take place from the SGSN to the mobile station or vice versa. For a real-time data service, minimum delay must be the first criterion in terms of quality. Hence, this is one very important KPI to be monitored in the optimisation process.

Network Performance Monitoring

The data can be collected from drive test results and the network management system (NMS). The process is quite similar to that explained in Chapter 2. Traffic monitoring will be key to optimisation of the network. Both the CS and PS traffic should be monitored, as the alignment of the GSM and GPRS traffic plays an important role in overall network quality (see Figure 5.9). The call voice KPI (mentioned in Chapter 2) and the data performance indicators could be used to monitor the traffic for a substantial time.

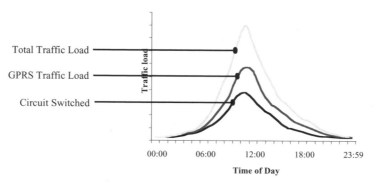

Figure 5.9 Example of traffic monitoring

Network Performance Assessment

As for a GSM radio network, assessment of a GPRS network is in terms of coverage, capacity, quality and related parameters.

Coverage
Drive tests results are used for assessment of coverage in a GPRS network. For the PS traffic, different coding schemes from CS-1 to CS-4 are used. As the coverage in a GPRS network is limited by interference, measurements are made under different propagation conditions and regional types (urban, rural, etc.) so as to get more accurate C/I criteria for planning coverage.

Initial values are based on the link budget calculation in the downlink direction (as it is assumed that the uplink power budget is equal to the downlink power budget). Also, in the initial network design the C/I and S/N values are considered to be similar. However, in a practical network they are usually different, so the actual coverage region may be quite different from the theoretical one. Thus, measurements and modifications leading to more realistic values of C/I and S/N are necessary.

Degradation of C/I values (and therefore coverage) may be due to multiple factors such as weak signal power (possibly due to propagation conditions), huge interference problems due to repeaters or an external source, or even poor frequency planning. Whatever the reason, insufficient cell coverage may lead to 'blind zones' or 'overlapping' in the radio network. The condition can be improved by increasing/decreasing the power of the BTS antennas (i.e. by adjusting the power budget) or by adjusting the cell selection parameters.

Capacity

A 'capacity bottleneck' in a GPRS network will most often be at the air- and G_b interface. These two interfaces will always have a direct impact on the system capacity and hence its quality. At the air-interface, the number of PDCHs will have an impact on capacity, so if capacity limitation is causing quality degradation then the number of PDCHs should be increased in the appropriate locations (e.g. hot spots, or areas where traffic increases during the busy hour).

The PCU also directly affects the allocation of radio resources. Whenever the PCU is short of capacity, TBF blocking will take place. Reduced throughput per time slot can be due to an increase in interference levels, or to the sharing of a time slot between several users. Thus, TSL capacity analysis becomes a part of the optimisation process. The number of TSLs defined for voice and packet traffic will need to be reviewed. It is possible that, during the launch of the network, the first TSL is a dedicated one while the remainder are for voice traffic. When the PS data traffic increases, the number of TSLs required for data also increases.

Moreover, the RLC protocol performance and the interference levels affect the capacity of individual TSLs. Re-transmissions take place in the data flow at the air interface. As the loading of the network (i.e. the number of subscribers on one TSL) increases, re-transmissions are reduced, so degrading the quality of the network. With an increased number of subscribers, the TSL capacity decreases due to the higher interference levels.

Quality

Network coverage, capacity and quality are of course interrelated. There may be low compatibility between the MS and network, there may be higher interference and degraded signal quality. Unavailable resources may cause problems with cell re-selection, channel unavailability, and packet data protocol activation.

Another important aspect in a GPRS network is the type of packet data and the desired QoS for it. When a GPRS network is launched, quality of service is an unknown quantity because of the huge variation in data types and its required quality. QoS has to be user-specific, which is generally stated in terms of data throughput and the associated delay. The quality required for individual services needs to studied, and solutions for critical ones need to be found and implemented. Most critical are real-time data, for which high capacity and minimal delay would be the solution.

Related Parameters

Parameters form the basis of network design and optimisation. Parameters need to be fine-tuned. The main parameters in this regard are those related to power control (coverage), GPRS territory (capacity), MS attachment success rate (quality), and cell re-selection.

5.6.2 Transmission Network Optimisation

Optimisation of the transmission network is quite similar to that for a GSM network. There are two important aspects that might come up in a heavily loaded GPRS network.

First, since the GPRS part of the network (packet core) may be added to an existing GSM network, there may be no changes required during the launch phase as there is low traffic. However, once the packet data users start to increase, the capacity of the existing transmission network is likely to need to be upgraded.

Second, the coding schemes in the GPRS network need to be reviewed. Usually CS-1 and CS-2 are used during the launch phase. If CS-3 and CS-4 are launched during optimisation or at a later phase, the A_{bis} time slot allocation will be changed.

5.6.3 Core Network Optimisation

Packet-core optimisation focuses on four issues: PCU/BSC and SGSN capacity, PDP functioning, and overall network security.

Generally the SGSN capacity is not an issue. However, as port requirements for the G_b interface increase, then it is possible that a new SGSN will be required. However, based on the criteria and limitations mentioned earlier in the section on TNP in a GPRS network, there is a possibility that the number of PCUs will have to be increased.

There may be problems related to PDP functionalities. This may be due to incompatibilities between different network elements or signalling such as HLR and SGSN or signalling between SGSN and GGSN (GTP signalling). Another problem can be congestion on the G_b interface due to a sudden increase in data traffic. The G_b interface might need to be readjusted for such a scenario.

Security is another issue. Owing to variation in the types of data being exchanged through the GGSN, the measures against security infiltration may need to reviewed.

6

EDGE: Network Planning and Optimisation

6.1 INTRODUCTION

GPRS networks are able to handle higher bit rates than GSM networks, but the data rates still fall short of what is required to make existing GSM networks deliver services at a speed comparable to that promised by third-generation networks The delay in the deployment of third-generation systems led to the emergence of a technology known as EDGE. This was capable of delivering services similar to those of third-generation networks, yet with implementation on the existing second-generation networks (e.g. GSM).

EDGE stands for 'enhanced data rates for GSM evolution'. The enhancement from GSM was to GPRS (i.e. voice and packet0, as covered in Chapter 5, while further enhancement of GPRS led to EDGE networks, as shown in Figure 6.1. The fundamental concept remains the same, i.e. voice, CS data and PS data being carried, and the network architecture is the same as in a GPRS network. Enhancement of HSCSD is known as ECSD (enhanced circuit-switched data), while enhancement of GPRS is known as EGPRS.

EGPRS implementation has a major effect on protocol structure (e.g. on layer 1 or layer 2). The modulation and coding schemes are quite different in EGPRS (this is explained later in the chapter).

In ECSD, though user data rates do not go beyond 64 kbps, fewer time slots are required to achieve this compared HSCSD. The architecture of ECSD is based on HSCSD transmission and signalling, thus having minimal impact on existing specifications.

In this chapter we will focus on EDGE network planning aspects from the EGPRS perspective.

Fundamentals of Cellular Network Planning & Optimisation A.R. Mishra.
© 2004 John Wiley & Sons, Ltd. ISBN: 0-470-86267-X

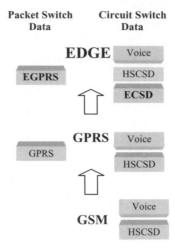

Figure 6.1 EDGE evolution

6.2 THE EDGE SYSTEM

As shown in Figure 6.2, the EDGE system is quite similar to the GPRS system (compare with Figure 5.1), but with the capability for higher data rates. The most important change is the new modulation scheme. In GSM and GPRS, the GMSK modulation scheme was used. In GMSK modulation, only one bit per symbol is used. In an EDGE network, octagonal phase-shift keying (8-PSK) modulation is used which enables a threefold higher gross data rate of 59.2 kbps per radio time slot by transmitting three bits per symbol. GMSK is a constant-amplitude modulation while 8-PSK has variations in the amplitude. This amplitude variation changes the radio performance characteristics, so hardware changes in the base stations are mandatory.

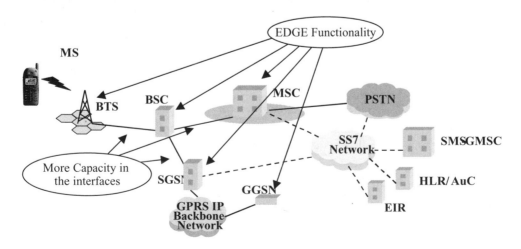

Figure 6.2 EDGE network (simplified)

6.3 EDGE NETWORK PLANNING

The radio network planning process is similar to that for GSM and GPRS networks. However, because of minor hardware and software changes in the existing network, that lead to major changes in network performance, the network planning parameters change quite a bit. Owing to increased bit rates, transmission planning undergoes a major change with introduction of the concept of a dynamic A_{bis} interface. In contrast, core network planning is virtually the same as for a GPRS network.

6.3.1 Radio Network Planning

Coding Schemes

As already mentioned, the EDGE system is an enhancement to the existing GPRS system. There are nine modulation and coding schemes (MCS-1 to MCS-9) that provide different throughputs, as shown in Table 6.1. The MCS scheme carries data from 8.8 kbps to 59.2 kbps on the existing GSM carrier of 270.833 kbps. For coding schemes MCS-1 to MCS-4, modulation is still GMSK; for MCS-5 to MCS-9 it is 8-PSK.

Based on this coding scheme, a data rate of 473 kbps (8 bits × 59.2 kbps) can be achieved. Though GMSK is a more robust scheme, 8-PSK gives more data throughput. However, the increased data rate comes at the price of decreased sensitivity of the system. This has an impact on network planning.

Another advantage in EDGE networks is that the switching between different coding schemes can take place easily, which was not possible in a GPRS network. When data transmission takes place in a GPRS coding scheme, it is not possible to switch the coding scheme on reception failure, so the re-transmission takes place with exactly the same protection as for its initial transmission. In EGPRS, it is possible to change the MCS, i.e. the data block can be sent again but with better protection than for its initial transmission. This is done through a process called link adaptation.

One more observation on the coding scheme is that the new set of GMSK coding is being used, e.g. CS-1 is equivalent to 8 kbps in GPRS networks (Chapter 5), with similar changes in coding schemes CS-2, CS-3 and CS-4. These changes are to provide incremental redundancy support, as will now be explained.

Table 6.1   Modulation and Coding Scheme in EDGE

MCS	Modulation	User rate (kbps)
1	GMSK	8.8
2	GMSK	11.2
3	GMSK	14.8
4	GMSK	17.6
5	8-PSK	22.4
6	8-PSK	29.6
7	8-PSK	44.8
8	8-PSK	54.4
9	8-PSK	59.2

Link Adaptation and Incremental Redundancy

As the propagation conditions change with time and region, the quality of the signal changes. Because of this, the modulation and coding schemes change all the time. Link adaptation is used for maximising the throughput per channel and changing the coding scheme depending upon the channel conditions. Basically, this leads to provision of the highest throughput possible with the lowest amount of delay. This gives better link quality and makes EDGE a more efficient system. Link adaptation (LA) algorithms are responsible for link adaptation. These algorithms activate the LA feature based on bit-error probability (BEP) measurements.

Incremental redundancy (IR) improves the throughput and is done by automatically adapting the total amount of transmitted redundancy to the radio channel conditions. This is achieved using two techniques: ARQ (automatic repeat request) and FEC (forward error correction). In the GPRS system, when errors are detected in the RLC blocks, re-transmission is requested and provided until the correct information reaches the destination. FEC provides the redundant user information that is used by the receiver to correct errors caused by radio channel disturbances. However, in an EDGE system, not all the redundant information is sent immediately. Only a small amount is send at first. If decoding is then successful, this saves a lot of capacity; if decoding is unsuccessful, then the normal ARQ process takes place. Because the LA algorithm selects the amount of redundancy for each individual transmission, this process basically reduces the number of re-transmissions and subsequent delays. The re-transmission mechanism in EDGE is more efficient (than in GPRS networks) by virtue of the IR phenomenon.

LA operates always on a first-sent block or a re-transmitted block, and upon receiving BEP measurements will change the MCS according to the condition of the network. IR is a specific re-transmission algorithm also known as 'hybrid ARQ II' (it includes puncturing, storage and soft combining at the receiving end) that if enabled together with LA will allow a change in MCS within the family. In order to change MCS during re-transmission, LA should be enabled.

Note that ETSI specifications make IR mandatory only for the MS as receiving side, not for the BTS.

Channel Allocation

Channel allocation in EDGE networks is nearly the same as in GPRS networks. BCCH, PCH, RACH, AGCH are the signalling channels, while PACCH is the only associated channel when the physical resources are assigned. Channel allocation algorithms are responsible for assignment of the channels to mobile stations. The EDGE base station should be capable of being synchronised with the existing GSM base stations. This will maximise the efficiency as both the base stations can be configured as one sector instead of two, thereby making only one BCCH necessary for the operation. Moreover, an increase in data rates leads to an increased signalling requirement for a given traffic and applications.

Smart Radio Concept

Implementation of the smart radio concept enhances the performance of the radio link, both in the uplink and downlink directions, by the use of diversity methods described later in the chapter.

6.3.2   Radio Network Planning Process

The basic process of radio network planning remains the same as in GPRS networks. However, because of the changes in the modulation and coding schemes, there are some changes in the coverage, capacity, and parameter planning of EDGE radio networks.

Coverage Planning

The link budget has a direct impact on coverage. As the EDGE network focuses more on PS data, the delay tolerance becomes a critical factor in defining system quality level. Some factors that affect link budget calculations specifically in EDGE are noted below.

Incremental Redundancy
Use of IR in coordination with LA not only makes re transmission more efficient but also optimises the performance of the system. It reduces the required C/I by at least 3 dB. Link budgets are calculated for a given modulation and coding scheme for a specific BLER. The BLER value affects directly the gain due to IR. In fact, the higher the BLER, the higher the IR gain.

Body Loss
When a mobile comes near to the human body, the signal level goes down. This is known as body loss. In GSM 900, this loss is typically 3 dB. No body loss is taken into account for packet data services in an EDGE network.

Diversity Effects
Use of diversity schemes generally has a positive impact on link performance, thereby increasing the area covered by individual sites. Both uplink and downlink diversity schemes are possible. However, smart radio concepts can be used to increase the coverage performance of an EDGE network drastically.

In uplink diversity, multiple antennas are used so as to cancel the correlated noise received at the antennas. Reduction of noise leads to a gain in the signal level. When there is no noise, the system allows the signal to flow without noise reduction. In the downlink direction, two transmitters are used and the signal is transmitted through two uncorrelated paths in bursts with slight delays. The transmitted power increases substantially with the use of two transmitters. Transmission of the signal over two different paths with delay reduces the effects of fast fading. Thus, the link performance and coverage can be increased substantially in EDGE radio networks as compared with GSM or GPRS radio networks.

Received Signal Strength
Signal strength in radio networks can be expressed in relation to interference signals. There are three parameters that specify this: E_b/N_0, E_s/N_0, and C/N. Both E_s/N_0 and C/N can be expressed in terms of E_b/N_0, which is the ratio of available bit energy to the noise spectral density (also known as the signal-to-noise ratio). E_s/N_0 is the ratio of energy-per-symbol to noise. For a GPRS system (with GMSK modulation) this ratio is unity, while for EDGE systems (8 PSK modulation) it is $E_b/N_0 + 4.77$ dB (as three bits is one symbol). C/N is

the ratio of total received power to total noise. The following are useful relationships:

$$E_b/N_0(dB) = E_s/N_0 - 4.77 \tag{6.1}$$

$$C/N(dB) = E_b/N_0 + 6.07. \tag{6.2}$$

Link budget calculations can be done by using MCS and BLER. However, received signal strength can be calculated for some specified data rates as well. E_s/N_0 can be used for throughput (per time slot) calculations in a cell. As seen in earlier chapters, the received signal strength is dependent upon the transmitted signal strength, losses, antenna heights and gains (TX and RX) and distance travelled by the signal. Loss in the signal strength gives the cell range. Thus, relationship between energy-per-symbol limited by noise can be plotted with respect to the cell range. Link level simulations with the above calculations can give the throughput per time slot for each coding scheme.

Capacity Planning

Capacity planning for EDGE networks is quite similar to that for GPRS networks, but the increase in throughput per radio time slot in EDGE changes some aspects of the planning. A brief overview of these concepts and their effects on throughput per radio time slot is given below.

The territory aspects explained in Chapter 5 (i.e. dedicated, default, CSW, etc.) stand the same for EDGE networks. Dedicated territory is specifically for PS traffic, CSW territory for CS traffic; default territory can be used for PS traffic if the CS traffic is not using it (CS traffic has a priority over PS traffic for default territory). The number of time slots that are assigned in each of these territories can be changed dynamically based on the load conditions.

The concept of frequency re-use is similar to that of GSM/GPRS radio networks. The re-use pattern defines the number of cells that can be used within a cluster in a manner such that no two neighbouring cells have the same frequency. A frequency re-use of 3/9 means that each frequency is used only once in three sites/cluster, wherein each site is three sectored. A frequency re-use of 1/3 will have a higher value of interference and thus would degrade the throughput per radio time slot. Thus, a higher frequency re-use value will give a higher throughput and less delay. However, the spectral efficiency is higher in cases of lower frequency re-use as fewer frequencies are being used. Time-slot capacities have a larger dynamic range compared with GPRS radio networks. The number of time slots available is fewer than the number of users in a cell (or network). This means that several users will be using same time slot, reducing the throughput per user. Thus, the higher the number of users (per time slot), the lower will be the throughput (per user) and the higher the delay. PS traffic can be allocated to the BCCH TRX or non-BCCH TRX. Since the spectrum efficiency is usually the same for BCCH and non-BCCH cases, the TSL capacity remains constant on the BCCH layer, making the BCCH layer more suitable for achieving high throughput. Usually, the frequency re-use patterns have been found to be more stringent on non-BCCH TRXs, hence BCCH TRXs are better at giving higher throughput. Enabling of frequency hopping does not have a major impact on capacity or quality of the EGPRS radio network.

Capacity planning dimensioning requires inputs related to cell configuration and traffic behaviour. This mainly includes the number of transceivers, definition of EDGE territories (number of time slots in CSW, dedicated and default territories), CS and PS traffic, etc. The outputs of capacity planning mainly include the amount of PS traffic, maximum, minimum and average PS load, available time slots of CS traffic, blocking rate, etc. Capacity planning based on this dimensioning is done on a cell basis to make sure that required capacity is available for CS and PS traffic, apart from signalling. Traffic types determine the signalling needs. Unlike in GPRS networks where short messages increase the resource (PRACH/PAGCH) requirements for channel set-up, EDGE networks do not face such a problem because of enhanced data rates. However, in EDGE networks, signalling requirements may be greater. This is because, if more and more users are able to get attached to the network using the same time slot even when the net load is constant (i.e. more users per time slot means less throughput per user), a decrease in TCH utilisation would take place during a TBF. If this TCH utilisation remains constant, more signalling channels are required, reducing the number of users getting connected to the network.

Thus, the steps involved in the dimensioning of an EDGE radio network can be summarised as follows:

- CS and PS territories are identified. For the PS territory, the number of time slots in the default and dedicated modes should be defined.

- Total traffic load inclusive of CS and PS should be defined.

- Delay versus load factor should be studied.

- Rate reduction factor/parameter must be calculated.

- Number of TRXs required for supporting the traffic is calculated.

Outputs of dimensioning can be:

- number of time slots required for voice traffic

- number of time slots required for data traffic

- number of time slots in default and dedicated territories

- average and maximum PS load

- average and minimum user throughput

- number of TRXs required to support the above parameters.

Parameter Planning

Apart from the parameters that have been discussed in earlier chapters, there are few additional parameters in EDGE radio networks. The most important ones are related to link adaptation and incremental redundancy.

Once EGPRS has been enabled; the initial coding scheme is selected. LA parameters are dependent on the modulation and coding scheme selections, for both initial transmission and re-transmissions. Although the MCS selections are based on the BTS parameters, the MCS used for the transmission is based on the BLER limits. However, MCS for both transmissions and re-transmissions can be affected by the mobile station memory.

Parameters related to multi-BCF and common BCCH assume importance in an EDGE radio network. As mentioned before, EDGE-capable and non-EDGE-capable transceivers in a one sector can be configured to have only one BCCH. TBF parameter setting makes it possible for TBFs of GPRS and EDGE radio networks to be multiplexed dynamically on one time slot. However, this scenario should be avoided, as the performance suffers in both the uplink and downlink. In the UL, GPRS performance suffers owing to the large amount of 8-PSK re-transmissions taking place, while in the DL, it is due to the GMSK modulations being used where 8-PSK is carrying higher data rates for EDGE.

Parameters related to delay and throughput assume importance owing to higher subscriber expectations from an EDGE network.

6.3.3 Transmission Network Planning

The transmission network planning process for an EDGE network is similar to that described in Chapter 3. However, owing to the higher data rates, there is a new functionality called 'dynamic A_{bis}'[*]. Thus, the following concepts assume importance in the design of an EDGE transmission network:

- dynamic A_{bis}

- dimensioning of dynamic A_{bis}

- dimensioning of the PCU/BSC.

Chapter 5 has already covered some aspects related to dimensioning of the PCU/BSC, so here we focus on dynamic A_{bis} and its dimensioning.

A_{bis}

The interface between the base station and the BSC is known as the A_{bis} interface, as shown in Figure 6.3. In GSM/GPRS networks this interface is 'static'. As we saw in Chapter 3, in GSM/GPRS networks the transceiver channels are mapped onto the A_{bis} PCM time slots. Each TCH uses two bits of PCM frame, and these two bits together are known as PCM sub-time slots. This is 'static' in nature because each TRX reserves its full capacity from the A_{bis} interface constantly even if there are no active users in the air-interface. This makes dimensioning of the A_{bis} interface straightforward compared to that in EDGE networks.

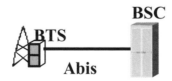

Figure 6.3 A_{bis} interface

Dynamic A_{bis} Functionality in EDGE Networks

Turning now to EDGE networks, octagonal phase-shift keying (8-PSK) changes data rates from as low as 8.8 kbps to 59.2 kbps, i.e. from coding scheme MCS-1 to MCS-9. Although voice signals are still carried in 16 kbps A_{bis} channels, for data this proves insufficient especially beyond coding scheme MCS-2. To carry more than 16 kbps traffic in the air-interface, the data traffic needs more that 16 kbps A_{bis} channels – probably 32, 48, 64 or 80 kbps. This data traffic is not there all the time, so the concept of 'dynamic' A_{bis} came into being. Figure 6.4 shows an example of 'static' and 'dynamic' A_{bis}.

---

[*] Dynamic A_{bis} feature is vendor specific. In equipments of some vendors, this feature may not be there. However, this feature clearly gives an edge to the EDGE transmission network as explained in this chapter.

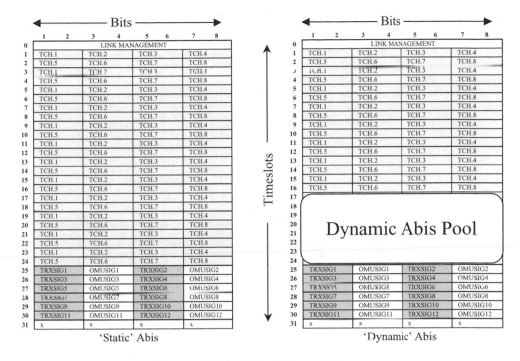

Figure 6.4 'Static' and 'dynamic' A_{bis}

A group of these dynamic A_{bis} channels is known as a 'dynamic A_{bis} pool' (DAP). A DAP consists of a minimum of one time slot, while the maximum may depend on the system capability within one PCM.

With EDGE transceivers, the BSC allocates A_{bis} capacity for data calls from the EGPRS dynamic pool (EDAP) when needed, i.e. when MCS-3 or higher is used. Standard GPRS (non-EGPRS) calls using CS-2, CS-3 and CS-4 can also use EDAP resources when allocated into EGPRS territory (EDGE transceivers).

Time-slot Allocation in Dynamic A_{bis}

Time-slot allocation in the A_{bis} interface of an EDGE transmission network is almost the same as in GSM/GPRS networks. Traffic channels on the A_{bis} still occupy 16 kbps sub-time slots for voice, and data rates up to 16 kbps. However, for data rates above 16 kbps, one master sub-time slot of 16 kbps and up to four sub-time slots from the DAP are required, as shown in Figure 6.5. The requirements for signalling channels for TRXs and BCF are the same as in GSM/GPRS networks. The pilot bits for synchronisation can be accommodated in the TS0 or TS31 (or any other time slot).

Dimensioning of the Dynamic A_{bis} Interface

The dynamic nature of the A_{bis} interface, and the allocation of a group of time slots for the DAP, changes the static capacity calculations of GSM/GPRS networks. In the latter, the A_{bis} interface allocates the whole capacity for the transceiver irrespective of usage; that

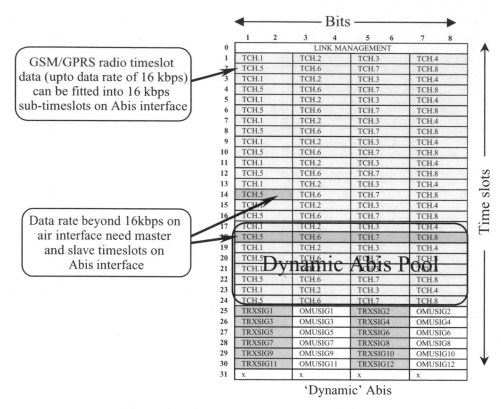

Figure 6.5 Timeslot allocation on dynamic A_{bis}

makes the capacity calculations simpler, because knowledge of the number of TRXs and the signalling rate leads to the amount of capacity required on the A_{bis} interface.

Before we look into DAP dimensioning, it should be remembered that DAP is used only for packet data for an MCS scheme greater than 2. Also, for each of the coding schemes, the number of required traffic channels will vary from none in MCS-1 (or CS-1) to four in MCS-8/9, as shown in Figure 6.6. (Designers should refer to specific product information and the system capabilities before they design the pool.) Also, the number of A_{bis} channels will increase when signalling is added to it, which again depends upon the system being used in the network. The A_{bis} channels that are present in the pool are also known as 'slaves'.

Based on the requirements of mobile subscribers, the sub-time slots from the pool are requested. The number of time slots allocated to the pool is usually in whole numbers – one, two, three, four or more time slots – depending on the DAP handling capacity of the system. Obviously, the number of time slots allocated to the DAP cannot be more than 30 (if TS0 and TS31 are 'reserved' for management purposes).

Dimensioning of dynamic A_{bis} means dimensioning of the pool. Once the number of time slots required for DAP is known, the remaining process becomes similar to dimensioning/planning of GSM/GPRS transmission networks.

The main idea behind DAP dimensioning is to find the number of time slots that can be assigned to a pool in such a manner that the A_{bis} interface does not become a limitation for

Coding Scheme	Bit rate (bps)	Minimum Abis PCM allocation
CS-1	8,000	◼
CS-2	12,000	◼
CS-3	14,400	◼
CS-4	20,000	◼ ◻
MCS-1	8,800	◼
MCS-2	11,200	◼
MCS-3	14,800	◼
MCS-4	17,600	◼ ◻
MCS-5	22,400	◼ ◻
MCS-6	29,600	◼ ◻
MCS-7	44,800	◼ ◻
MCS-8	54,400	◼ ◻◻◻◻
MCS-9	59,200	◼ ◻◻◻◻

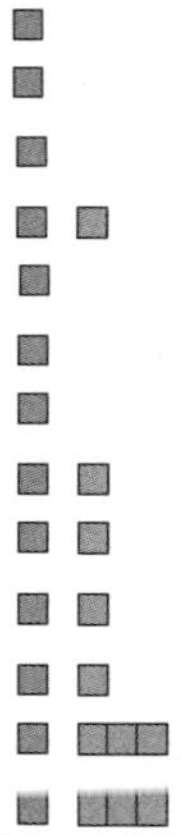

Figure 6.6 Transmission requirement for EGPRS coding schemes at the A_{bis} interface

the air interface throughput. Thus the main inputs for this would be the number of radio time slots required for PS traffic (i.e. dedicated and default territory in an EDGE radio network), the capacity of radio time slots, blocking that could take place on the A_{bis} interface, etc. Based on these inputs and the number of PCM time slots available for the pool, the required number of PCM time slots can be calculated. Obviously, limitations and capability of the equipment needs to be taken into consideration. So:

$$B(n, N, p) = \sum_{x=N=1}^{n} P(x) \qquad (6.3)$$

where B is the blocking probability of the DAP, N is the number of time slots available in the pool, p is the utilisation of EDGE channels in the air interface, and n is the number of traffic channels used in the air interface.

This dimensioning will result in outputs such as the maximum throughput possible on the air-interface with A_{bis} acting as a bottleneck, and the number of time slots required to allow maximum possible throughput on the air interface to be carried on the A_{bis}.

The Impact of Dynamic A_{bis} on Transmission Network Design

Assume that there is a base station site having a configuration of $3 + 3 + 3$. When operating in the GSM mode, using 16 kbps signalling, the required number of PCM time slots per transceiver is 2.25 (two for traffic and 0.25 for 16 kbps signalling). This means that the total number of time slots required for these nine TRXs is $9 \times 2.25 = 20.25$; i.e 21 time slots approximately. This would mean that, of 32 time slots, TS0 and TS31 along with these 21 are used (23 in total), leaving nine time slots available for possible future upgrades. In an EDGE network, these nine time slots might be used for the DAP, if they are sufficient.

Now consider this EDGE network using the MCS-9 coding scheme; i.e. each user logging on the network for PS services requires four A_{bis} channels, or one full additional time slot, apart from the master sub-time slot on the PCM. This means that at a given time only nine users can use the system at a maximum rate of 59.2 kbps. As soon as a tenth subscriber logs on to the network also requesting the MCS-9 scheme, the system starts sharing its resources. This means that although access will be given to this tenth subscriber, the request will be downgraded to lower coding schemes like MCS-6 or MCS-7. Also, the coding scheme of the existing subscribers logged on may be downgraded to MCS-6 or MCS-7, in order to accommodate this new subscriber. So the throughput on the air-interface goes down because of the limitation on the A_{bis} interface. In this case it would be advisable to have more time slots in the pool, implying more PCMs required per site, with additional cost. Thus, the number of time slots planned for DAP should be a balance between the desirable throughput and cost.

Example 1: Dynamic A_{bis} Dimensioning for One Single Site
Consider a single site with configuration $1 + 1 + 1$. It should be able to deliver a data rate of 59.2 kbps (i.e. MCS-9) by using two radio time slots. What is the number of time slots that should be reserved for dynamic A_{bis} so that the A_{bis} data rate reduction factor is less than 1%; and how many E1s (PCMs) are required for the BTS–BSC connection? (PCMs are known as E1s (in ETSI) and T1s (in ANSI) standards.) Signalling is on the GSM layer.
*   As one user is able to use two radio time slots to get a data rate of 59.2 kbps, four users can use eight time slots at a given time. This implies 4 data Erlangs.*

$EGPRS\ territory = 8$

$Data\ rate = 59.2\ kbps$

$Data\ rate\ after\ C/I = 40\ kbps$

$For\ a\ DAP = 12\ TSL\ on\ A_{bis}$

Based on the binomial distribution formula stated, the results are as follows:

- The sharing probability of the DAP is only 3.176%.

- The A_{bis} data rate reduction is less than 1%, which means that the A_{bis} will not be a bottleneck for the throughput of the air-interface.

- The average throughput per RTSL, limited by radio and A_{bis}, is 19.88021.

- The A_{bis} data rate reduction is 0.60%.

Now, as three transceivers need only $3 \times 2.25 = 6.75$ time slots, and the DAP will require 12 time slots, the total number of time slots needed will be $6.75 + 12 = 18.75$. Apart from this, TS0 and TS31 can be kept for management, etc. Thus, one E1 will be sufficient for connecting this site to the BSC.

6.3.4 Example of RNP + TNP Dimensioning

An example of the combined result of radio and transmission network dimensioning is shown in Table 6.2. This is the result of a dimensioning case where the maximum throughput per user is 96 kbps, but owing to C/I limitation the acceptable user throughput is reduced to 64 kbps per user. The dimensioning is done for coding scheme MCS-7. The EDAP size is nine time slots.

Table 6.2   RNP + TNP Dimensioning output

1	Data Traffic in the BH	DT		96	Kbit/s per cell
2	Acceptable avergage user throughput			64	Kbit/s
3	RTSL per mobile	Nu		2	
4	Timeslot Capacity	k		48	Kbit/s
5	PS Traffic intensity	Tps	DT / k	2	Data Erlangs
6	Number of RTSL per cell	Ns		4	
7	Utilization		Tps / Ns	0.5	
8	Reduction Factor f (Nu = 2, Ns = 4, Utilisation)	RF	Calculated by RF	0.75	
9	Avg user throughput (without transmission limitations)	Trg	RF*Nu*k	72	
10	**EDAP pool size**			**9**	
11	**Abis data rate reduction factor**	**AbisRF**	**Calculated by TRS**	**0.9962**	
12	**Avg user throughput (with Abis transmission limitations)**	**E**	**Trg * AbisRF**	**71.728**	

6.3.5   Core Network Planning

Both CS and PS core network planning remain similar to what was described in Chapters 4 and 5. The major change in the packet core section is enhanced capacities of the BSC and SGSN to cater to higher data rates on the radio network. Dimensioning of the packet core network mainly involves the PCU, the G_b interface and the SGSN, as follows:

- PCU dimensioning:
  Number of transceivers that can be supported by one PCU
  Amount of (data) traffic that can be handled by one PCU
  Number of traffic channels that can be handled by one PCU

- G_b interface dimensioning:
  Minimum number of G_b interfaces required between the PCU and SGSN
  Uplink and downlink traffic calculations for the frame relay links

- SGSN dimensioning:
  Number of subscribers
  Processing capacity of the SGSN
  G_b interface capacity.

6.4   NETWORK OPTIMISATION

Optimization of an EDGE network will focus mainly on the radio and transmission networks owing to the increased data rate on the air-interface and the introduction of the dynamic A_{bis} feature on the A_{bis} interface. Usually, optimisation of radio and transmission networks is performed separately, but this scenario changes owing to the fact that the dynamic A_{bis} pool, if not properly planned, may become a bottleneck for the air interface throughput.

There are three main steps, as shown in Figure 6.7. The 'conventional' radio network optimisation process will be quite similar to that explained in earlier chapters for GSM/GPRS

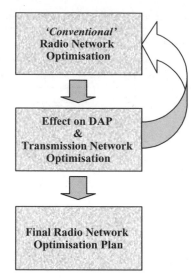

Figure 6.7 Radio network optimisation in EDGE network

networks. Transmission network optimisation remains similar also, except for optimising the A_{bis} capacity so that it does not become a bottleneck for the air-interface.

6.4.1 Radio Network Optimisation

Coverage, capacity and quality can be improved by using the standard principles of optimisation that have been described earlier for GSM/GPRS. The main focus in EDGE radio network optimisation will be on throughput, which affects the capacity and coverage directly. One important point to remember is that, since EDGE implementation will generally be on top of an existing GPRS network, it is essential that throughput improvement is measured and analysed with respect to that of the GPRS network.

6.4.2 Transmission Network Optimisation

The optimisation process is similar to that in GSM networks, however, there are two extra factors that need to be considered during the EDGE optimisation process: the dynamic nature of the A_{bis} interface, and the air-interface throughput changes due to the bottleneck created by the A_{bis}. As there is a pool of time slots reserved for packet data, it will always be the case that the number of packet service users increases in the network. Though the pool itself will be dynamic, the number of time slots that will be used for the pool will be static; so once it is figured out that the required time slots (in the pool) need to be increased, some optimisation will be required. The situation is more severe if the number of PCMs required per base station also needs to be increased.

Key Performance Indicators

The major extra KPIs relate to the definitions of the EDGE territory, BLER (block error rate), MCS schemes used, BCCH usage, etc. These parameter settings define the initial throughput. Adjustments to these settings (apart from those in the GSM/GPRS network) will permit maximisation of throughput.

Performance Measurements

As usual, performance measurements involve drive tests, reports from the network management system (NMS), and network planning tools. However, the focus will be on throughput in an EDGE network, and the key areas of performance measurement will therefore be at the air interface and the A_{bis} interface. For the air interface, the interest will be on capacity, coverage and quality; while for the A_{bis} interface it will be on the variation of throughput with respect to modulation and coding schemes. One such example of performance based on accessibility, retainability, quality (ARQ) and peak traffic is shown in Figure 6.8.

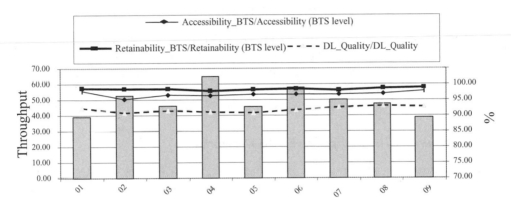

Figure 6.8 Test results (ARQ and peak traffic: BTS 1)

As stated in earlier chapters, performance indicators like the ability of mobile stations to get access to the cell (accessibility), DCR (retainability), call quality in the downlink (DL-quality), etc., need to be measured. Figure 6.9 shows an example of BTS performance measuements.

Date	BTS ID	Accessibility (BTS level)	Retainability (BTS level)	(dB) UL_Quality	(dB) DL_Quality	TCH Erlang
04/06/03	BTS1	98.61	99.05	92.73	72.01	0.55
04/07/03	BTS1	98.21	96.49	91.35	91.23	0.68
04/08/03	BTS1	66.18	100.00	99.11	99.33	2.06
04/09/03	BTS1			100.00	95.62	0.11

Figure 6.9 BTS performance measurements

Measurement of the A_{bis} throughput capacity with respect to the coding schemes used by mobile subscribers will be critical in adjusting the boundaries of the dynamic pool. The example in Figure 6.10 shows the variation of the pool with respect to modulation and coding schemes.

Performance of an EDGE network will usually be in terms of the underlying GPRS network. The achieved mean throughput values and delay values in an EDGE network are

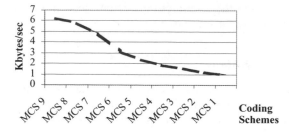

Figure 6.10 MCS versus throughput

better than for the GPRS network, even when the load is higher. The measurement information serves not only to increase coverage and capacity, but also the quality of service experienced by end-users, which will be closely observed in EDGE networks. The optimum value of QoS will depend ultimately on traffic requirements. As will be seen in Chapter 7, interactive traffic has more stringent quality requirements than background traffic. Spectrum efficiency will play a big role in achieving this. The tradeoff between the number of transceivers and interference levels (i.e. between higher throughput and more users) is a delicate issue. Final optimisation plans will be similar to those for GSM/GPRS networks, but with stringent QoS requirements guiding the results.

Improvement of Throughput in EDGE Networks

QoS depends on the service required, but higher throughput will be the main QoS-fulfilling requirement in an EDGE network. Throughput on the air interface and coverage (both at hot

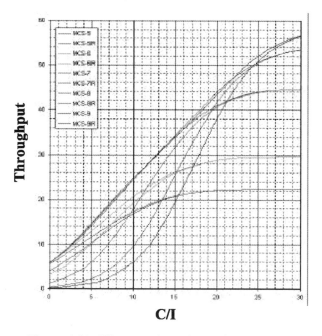

Figure 6.11 C/I versus throughput (simulations)

spots and cell edges) will be the main focus in the planning as well as optimisation of the radio network. Throughput is dependent upon a number of factors such as C/I, handovers, MCS changes, etc.

An increase in the C/I ratio will result in increased throughput (shown in the simulation in Figure 6.11). Measurement at the cell edges will be critical, and here performance improvement can be achieved by using concepts such as antenna tilts, power increase/decrease, etc.

Increasing the number of handovers will also decrease throughput, so optimisation of the neighbours to reduce unnecessary handovers can be one possible solution to increase throughput. Throughput also decreases with the number of MCS changes from higher-throughput MCS to lower-throughput MCS. Optimisation can be achieved by parameter auditing after observing the C/I and receiver level statistics against MCS changes.

Conclusions

Traffic loads (both CS and PS), and the applications used, affect the performance of an EDGE network. C/I criteria are stricter than in GPRS networks. The spectrum efficiency in an EDGE radio network is higher than in GPRS networks, which is mainly because of the higher throughout and better link performance. Though the performance measurements and subsequent results are usually based on data throughput, in real network optimisations both voice and data traffic performance have to be assessed. With numerous types of application running on the network, there will be a lot of interaction amongst the applications. The performance indicators noted in earlier chapters for radio/transmission/core networks, and their interactions, are a key focus in the optimisation of an EDGE network.

III

Third-generation Network Planning and Optimisation (WCDMA)

7

3G Radio Network Planning and Optimisation

Third-generation UMTS (universal terrestrial mobile system) networks have a predominance of data traffic, unlike GSM networks. The rate at which this data traffic can move will be significantly higher than that offered by GSM/GPRS/EDGE networks. For this reason, the third-generation UMTS networks are fundamentally different from the existing GSM systems.

7.1 BASICS OF RADIO NETWORK PLANNING

We have seen some basics of third-generation networks based on the WCDMA technology in Chapter 1. As in GSM networka, radio network planning takes an important place in the planning process owing to the proximity to the subscribers.

7.1.1 Scope of Radio Network Planning

As noted in Chapter 1, the radio and transmission networks have been combined to form the 'radio access network' (RAN), but for ease of study we will liik here into the scope of the radio network planning engineer's work. The focus area is quite similar to that of the GSM/GPRS networks, as shown in Figure 7.1, that is, planning capacity and coverage for an upcoming third-generation network having a desired quality.

Before we go into the details of radio network planning and optimisation in 3G networks, let us look at a few concepts that will help in an understanding of the planning process.

7.1.2 System Requirements

3G networks serve a different purpose to earlier networks, and major changes from previous network types are:

Fundamentals of Cellular Network Planning & Optimisation A.R. Mishra.
© 2004 John Wiley & Sons, Ltd. ISBN: 0-470-86267-X

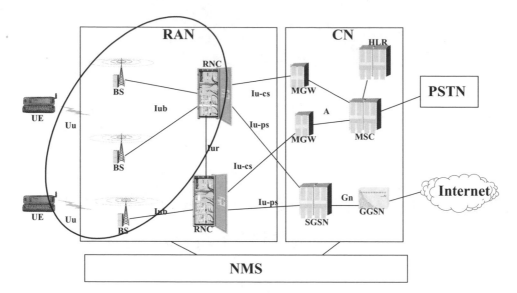

Figure 7.1 Scope of radio network planning in a 3G system (WCDMA)

- maximum user bit rates up to 384 kbps

- efficient handover between different operators and technologies (e.g. GSM and UMTS)

- an ability to deliver requested bandwidth

- an ability to deliver different services (both CS and PS) with the required quality.

7.1.3 WCDMA Radio Fundamentals

Let us try to understand some principle that would be of use in designing a network capable of giving the desired performance. WCDMA (wide-band code-division multiple access) technology has emerged as the preferred and most adopted technology for the third-generation air interface. Air interface technology has changed, so there are some major differences between the WCDMA and GSM air interfaces:

- The WCDMA system supports higher bit rates, so a large bandwidth of 5 MHz is used as compared to 200 kHz in GSM.

- Packet data scheduling in WCDMA is load-based, while in GSM/GPRS it is time-slot based.

- Theoretically, only one frequency channel is used in WCDMA, while GSM uses many frequency channels.

- The limited bandwidth of 5 MHz is sufficient for radio network design. Multipath diversity is possible with rake receivers, while in GSM techniques like frequency hopping are used for (frequency) diversity.

- Quality control in WCDMA is done using RRM algorithms, while in GSM it was done by implementing various techniques such as frequency planning.

- Users/cells/channels are separated by codes instead of time or frequency.

7.1.4 Service Classes in UMTS

In a 3G network, mobile equipment will be able to establish and maintain multiple connections simultaneously. The network will also allow efficient cooperation between applications with diverse quality of service (QoS) requirements, as well as adaptive applications that will function within a wide range of QoS settings. From the users' perspective, third-generation networks will be able to give high quality for many services. This means that all the sections of the network RAN and CN will be trying to achieve these quality standards (from the users' perspective and defined by the ITU). The quality can defined by two main parameters:

- guaranteed and maximum bit rate (kbps) possible
- permissible delays (ms).

Both single-media and multi-media services will be handled in the third-generation networks. Based on the QoS criteria, multi-media services have been further classified as:

- conversational
- streaming
- interactive
- background.

The *conversational class*, as the name suggests, is for applications like speech. It is the most delay-sensitive of the four classes. A typical example of this class is video telephony, voice-over-IP (VoIP). In this class the delay is based on the human perception of the application, hence has strict requirements for quality.

The *streaming class* refers to traffic flow that is steady and continuous. It is server-to-user type. The most common example in this class is the Internet. In 3G networks, the Internet will be faster because a user will be able to see data before it is completely downloaded. There are two components of this class: messaging and retrieval. A typical example is downloading of streaming videos (e.g. news).

Web browsing is a typical example of the *interactive class*. In this case the user requests data from a remote entity (e.g. a server). Location-based services are an example of this class. A user will be able to access information like bus and train timetables, flight schedules, and any local data that might be useful.

Short messages, file transfers, etc., come into the *background class*. Nearly all the traffic that does not fall under the first three categories are included, such as e-mails. This class of service has the least stringent quality of service requirement of all the four classes.

7.1.5 Elements is a WCDMA Radio Network

User Equipment (UE)

The mobile terminal is called user equipment. The basic principles remain the same as explained in Chapter 2, but with the addition of the capability to handle data calls. User equipment can be divided into three parts, USIM, ME and TE, as shown in Figure 7.2(a).

The USIM card (also known as SIM) contains authentication information and associated algorithms, encryptions and subscriber-related information. In contrast, the mobile

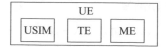

Figure 7.2(a) Simplified block diagram of user equipment (UE)

equipment (ME) is user-independent. The Terminal equipment (TE) is responsible for termination of the entire control and user-plane bearer with the help of the ME.

Base Station (BS)

The base station is also known as 'node B' in a WCDMA radio network. It is more complex than the base station of a GSM network. Its functions include handover channel management, base-band conversion (TX and RX), channel encoding and decoding, interfacing to other network elements, etc. A simplified version of it is shown in Figure 7.2(b).

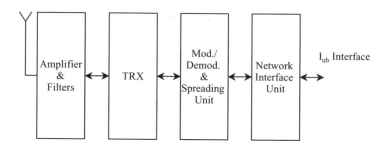

Figure 7.2(b) Simplified block diagram of WCDMA base station

Amplifiers and Filters
This unit consists of signal amplifiers and antenna filters. The amplifiers are used to amplify the signal coming from the transceiver and going towards the RF antenna (the downlink signal), while the filters select the required frequencies coming in from the RF antenna (the uplink signal) and amplify the signals for further processing before sending them to the receiver part of the TRX.

Transceiver
The TRX is capable of transmitting and receiving signals, by handling uplink and downlink traffic. It consists of one transmitter and one or more receiver.

Modulation/Demodulation and Spreading Unit
This unit is responsible for modulating the signal in the downlink direction and demodulating in the uplink direction. It is responsible for summing and multiplexing the signals and also processing the signals. This unit contains the digital signal processors that are responsible for coding and decoding signals.

Network Interface Unit

This unit acts as an interface between the BS and the transmission network or any other network element, such as co-sited cross-connect equipment.

Radio Network Controller (RNC)

The radio network controller (RNC) is similar to the BSC in GSM/GPRS networks, but is rather more complicated and has more interfaces to handle. The RNC performs radio resource and mobility management functions such as handovers, admission control, power control, load control, etc.

In fact the RNC plays a dual role in a WCDMA radio network, which should be understood from a network planning perspective. A radio network controller can be SRNC (serving RNC) or DRNC (drifting RNC). From one mobile, if the RNC terminates both the data and related signalling then it is called the serving RNC. If the cell that is used by this UE is controlled by an RNC other than the SRNC, then it is called the DRNC.

More details on the structure of the RNC are given in Chapter 8, in the discussion of access transmission in UMTS networks.

7.2 RADIO INTERFACE PROTOCOL ARCHITECTURE

7.2.1 Introduction

Based on the OSI reference model, Figure 7.3(a) shows the first three layers of the WCDMA radio interface protocols. These three layers are needed for the functioning (set-up, release, configuration) of the radio network bearer services.

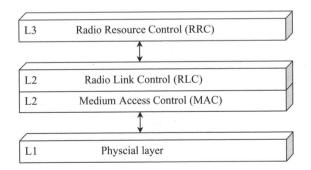

Figure 7.3(a) Basic radio interface protocol architecture

Layer 1 is the physical layer, the actual medium of transfer. Planning engineers should note that this layer is not just a 'physical medium' but should also be able to perform certain functions. The main functions of layer 1 include RF processing, modulation/demodulation of the physical channels, multiplexing/demultiplexing of the physical channels, error detection and correction, rate matching, power control, synchronisation, etc.

Layer 2 is the link layer. It is required because of the need to allocate minimum resources for a constantly changing data rate. It has two main sub-layers within itself: RLC and MAC.

There are two other layers – called the 'packet-data convergence protocol' (PDCP) and 'broadcast – multicast control' (BMC) – but they exist only in the user plane.

The MAC (medium access control) layer in an entity that is responsible for the mapping of the logical channels to the transport channels. It provides data transfer services on the logical channels. As it is an interface between L1 and L3, it also provides functions like multiplexing and demultiplexing of packet data units to/from the physical layer. The MAC layer is also responsible for measurements related to traffic volume on the logical channels and further reporting to layer 3.

Functions like segmentation and reassembly of the variable-length packet data into smaller payload units is done by the RLC (radio link control) layer, a sub-layer of layer 2. Another important function of this sub-layer is error correction by re-transmission in an acknowledged data transfer mode. Other functions include controlling the rate of information flow, concatenation, ciphering, and preservation of the higher-order PDUs.

There are three modes of configuring an RLC by layer 3: transparent mode (no protocol overhead added), unacknowledged mode (no re-transmission protocol in used, so data delivery is not guaranteed), and acknowledged mode (a re-transmission protocol is used and data delivery is guaranteed). PDCP and BMC protocols exist only in the user place. PDCP is only for packet data, with its major function being compression of the PDUs at the transmitting end and decompression at the receiving end in all the three modes of operation, transparent, unacknowledged and acknowledged. BMC functions only in the transparent and unacknowledged modes, providing broadcast/multicast scheduling and transmission to the user data.

Layer 3 is also contains sub-layers, but the radio resource control (RRC) sub-layer is the one that interacts with layer 2. It handles the control plane signalling between the UE and network in connected mode. It is also responsible for bearer functions like establishment, release, maintenance and reconfiguration in the user plane and of radio resources in the control plane. Functions of the RRC include radio resource management and mobility management, as well as power control, ciphering, routing (of PDUs) and paging.

7.2.2 Protocol Structure for Universal Terrestrial Radio Access Network (UTRAN)

The protocol structure for UTRAN is based on the model described above and can be used further to study the protocol structure for different interfaces in detail. As shown in Figure 7.3(b), there are two main layers, the radio network and the transport network, and two planes, the user plane and the control plane.

The visible part of the network is the radio layer, while the transport layer elements (or equipments/technology) can vary without making any changes to the radio layer characteristics. Transportation of all user-specific data CS or PS is done through the user plane, while the transport plane is responsible for all signalling activities in the network. The control plane protocol present in the radio network layer is known as the 'application protocol'. It includes protocols such as RANAP, RNSAP and NBAP (see below). The transport network layer has another control plane known as the transport network control plane that is responsible for all the signalling within it. The protocol here is known as ALCAP. The transport layer has its own user plane also for the data bearers in the transport layer.

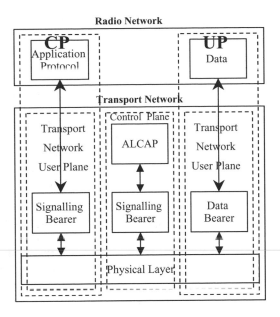

Figure 7.3(b) General protocol structure for UTRAN

- RANAP (radio-access network application protocol) is the signalling protocol defined between the RAN and CN. The main function of RANAP is to control the resources on the I_u interface, providing both control and dedicated control services.

- RNSAP (radio network sub-system application protocol) provides the signalling on the I_{ur} interface. As the I_{ur} interface is present between the two RNCs (one is SRNC and the other is DRNC), it is managed by these two RNCs. RNSAP is responsible for management of bearer signalling on the I_{ur} interface apart from the transport and traffic management functions.

- NBAP (node B application protocol) maintains the control plane signalling on the I_{ub} interface. This protocol is responsible for communications between the node B (WCDMA base station) and UE.

- ALCAP (access link control application part) is required for setting up of the data bearer and signalling in the transport network control plane. The presence of ALCAP is necessary as the user plane and the control plane can be separated and independent from each other. ALCAP may or may not exist depending upon the type of data bearer.

7.2.3   Channel Configuration in WCDMA radio network

As mentioned above, the MAC layer is responsible for mapping the logical channels on to the transport channels. It provides the data transfer services on the logical channels. As usual, logical channels are composed of control and traffic channels. These are again subdivided into common and dedicated channels. These channels along with their functions are noted in Table 7.1.

Table 7.1   Logical channels in a WCDMA radio network

Channel	Abbreviation	Function/application
Broadcast common control channel (DL)	BCCH	Transmits the system control information
Common control channel (UL/DL)	CCCH	Used (usually by the UE) for transmitting information related to control between network and UE
Common traffic channel (DL)	CTCH	Used to transmit dedicated user information to a group of UEs
Dedicated control channel (UL/DL)	DCCH	Dedicated channel for control-related information between the UE and network
Dedicated traffic channel (UL/DL)	DTCH	Similar to DCCH except that it is used for user information
Paging control channel (DL)	PCCH	Used to page information to the UE

7.3   THE SPREADING PHENOMENON

7.3.1   Introduction

The spreading phenomenon is the basis of the WCDMA technique. The word 'spreading' refers to the 'spreading of bandwidth' of the actual information. This means that the information is transmitted at a larger bandwidth than the original bandwidth. Through this process, the modulated signal becomes tolerant towards a narrow-band interfering signal. This solves the problem faced by TDMA and FDMA systems, which limit the simultaneous number of users. By using the spread spectrum technique, efficient use of the spectrum will allow multiple users to use the same frequency band. Using a very common technique known as *direct sequencing*, the information can be spread over a frequency spectrum. The most common techniques used for this purpose are:

- DS-WCDMA_FDD (direct-sequence WCDMA frequency-division duplex)

- DS-WCDMA_TDD (direct-sequence WCDMA time-division duplex)

- MC-CDMA (multi-carrier code-division duplex)

With DS-WCDMA_FDD, the information is spread over the frequency spectrum, while with DS-WCDMA_TDD the frequency band is located on both sides of the WCDMA_FDD signal. With MC-CDMA, the whole of the frequency band is used with a number of carriers instead of one. Both of the techniques FDD and TDD will be implemented, but in the initial phase of a third-generation system only the FDD mode is used because it is more suitable for outdoor coverage purposes. The frequencies for the DS-WCDMA_FDD are:

- downlink: 2110–2170 MHz

- uplink: 1920–1980 MHz.

As seen from these frequencies, the bandwidth is 60 MHz in either direction, while the separation between the uplink and downlink is 190 MHz.

7.3.2 Symbols and Chips

The phenomenon of spreading (and de-spreading) is shown in Figure 7.4. The signal is 'spread' by modulating the original signal (referred to as data in the figure). Usually, a BPSK (binary phase-shift keying) modulated signal is used as the original signal. This means that the signal modulation is performed a second time on the original signal. This original data signal (obtained after BPSK modulation) is then modulated by multiplying it by a sequence of bits (the wide-band spreading signal), thus converting the original BPSK narrowband signal into a spread signal of wider bandwidth. Each bit of the original signal (also known as a 'symbol') is multiplied by a sequence of bits called 'chips'. These chips are a part of the signal that is used to multiply the original data, the 'code' signal.

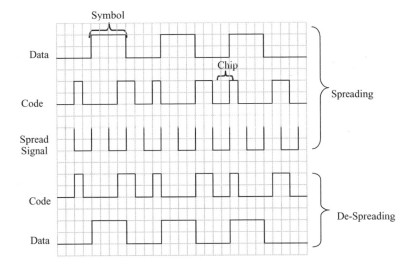

Figure 7.4 Spreading and de-spreading

The spread signal is then transmitted to the receiver. At the receiving end, the received signal is then again multiplied by the same code to retrieve the original data. The original data stream can be obtained only if the code signals are identical.

In the radio network, there are many subscribers transmitting and receiving signals at the same time. The spreading codes achieve separation between these signals. Here we can define the term 'spreading factor' (also known as 'processing gain'), which is the number of chips used by one symbol. Processing gain can also be defined as the ratio between the chip rate and the bearer bit rate. The processing gains, together with the wideband nature, suggest a frequency re-use of 1 between different cells/sectors. Thus, we can come up with some important information here:

- One WCDMA carrier is 5 MHz. With the guard band taking 1.16 MHz of bandwidth, the effective bandwidth is 3.86 MHz.

- One bit of the base-band signal is called a symbol.

- One bit of the code signal is called a chip. The chip rate is defined to be 3.84 Mcps (million chips per second). This means that one chip is 0.26 μs.

Now these spreading codes contain two more codes: scrambling code and channelisation code. Actually:

$$\text{Spreading code} = \text{scrambling code} \times \text{channelisation code}.$$

Scrambling Codes

Scrambling codes are used on top of the spreading codes, thus not changing the bandwidth of the signal but only making it different for the purpose of separation. These codes are used to separate the users and cells (and/or base station) in the uplink and downlink directions respectively.

In the downlink direction, it is possible to generate about 262 148 ($2^{18} - 1$) scrambling codes. However, not all these codes can be used. The scrambling codes are divided into 512 groups and each of these groups has one primary and 15 secondary scrambling codes. Based on these principles, it is possible to have 8191 codes.

In the uplink direction, there are 2^{24} scrambling codes. These uplink codes are subdivided into long and short codes.

Channelisation Codes

Channelisation codes are used in the downlink direction to separate users and channels within the same cell. In the uplink direction they are used to separate data and control channels from same user equipment. These codes are based on a technique known as OVSF (orthogonal variable spreading factor). The OVSF technique maintains the orthogonality between the different spreading codes while allowing the spreading factor to be changed. These codes increase the transmission bandwidth.

7.3.3 Rate Matching

This is a process by which the number of bits that are to be transmitted is matched to the number of bits that are available on a single frame. If the number of bits transmitted is lower than the maximum, then the transmission may get interrupted. Thus processes such as repetition and puncturing are used to make the two numbers (transmitted and maximum) equal.

In the uplink direction, *repetition* is the preferred way of rate matching. In the downlink direction, *puncturing* is the preferred method. UL rate matching is a dynamic process that varies with the processing of every frame. This process is important in 3G networks as it is one way in which desired QoS can be achieved by fine-tuning the bit rates.

7.4 MULTIPATH PROPAGATION

Multipath propagation is an interesting phenomenon in WCDMA radio networks. Multipath propagation takes place when a signal takes multiple paths and reaches the receiver. This means that the same signal reaches the receiver at different points of time, with different amplitudes and phases. This may lead to the phenomenon of fast fading (as described in Chapter 2), especially when the signals arrive at a difference of half a wavelength so that deep fades take place.

RAKE receivers are used in WCDMA radio networks, being a more efficient way to receive multipath signals. A RAKE receiver contains many receivers and is able to allocate them based on the timing of incoming signals, code generators and amplitude/phase detecting equipments. These receivers (also known as RAKE fingers) are able to track the fast-changing amplitudes and phases originating from the fast-fading process and remove them. Typically, RAKE is able to handle four fingers. The numbers of RAKE fingers used in the BS and UE are usually different.

7.5 RADIO NETWORK PLANNING PROCESS

The radio network planning process for WCDMA is nearly the same as for GSM networks. The main aspects include preplanning/dimensioning, fieldwork of site surveys, etc., and subsequent modification of the parameters based on the fieldwork, before the final plans lead to the commissioning phase. These aspects continue until the sites go on air. The quality of service (QoS) targets are more stringent in WCDMA radio networks compared with GSM radio networks. With each kind of service demanding a different QoS, planning of the network becomes a daunting task.

7.5.1 The Pre-planning Phase

The most important work in the pre-planning phase is dimensioning of the network based on inputs and assumptions for getting a desired coverage and capacity (see Figure 7.5).

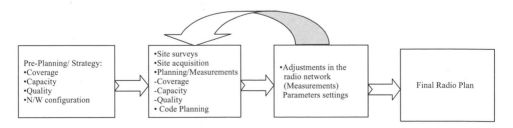

Figure 7.5 Radio network planning process

Coverage of the WCDMA radio network is dependent on conditions such as the expected area to be covered, the type of area, network configurations (and system capabilities) and propagation conditions. Defining the base station location is an important part of this process. The number of base stations required to achieve the desired coverage and quality is also dependent on the capacity of the base stations. A link budget calculation and propagation models form an important part of the coverage predictions.

Capacity planning in the initial phase is a major challenge. With so many types of applications having varying quality and capacity requirements, capacity and quality may prove to be a limitation, especially for real-time data service applications. The frequency availability, subscriber base and growth, and type of services required will be major inputs for the dimensioning exercise, and the main results would include the number of sectors and transceivers that would be required for these base stations.

As noted in Chapter 2, coverage, capacity and quality go hand-in-hand, and network configuration plays a big role in achieving the desired service standards. There are four different types of service classes and numerous different kinds of services in each class. Enough coverage, sufficient capacity and desired throughput will be key in the network configuration design to achieve the desired quality standards.

7.5.2 Structure and Performance of the Physical Layer

In a WCDMA radio network, the structure of the physical layer directly affects the performance of the network, so some understanding of this is necessary to dimension the network. We have already seen the logical channels in Table 7.1. Here we briefly look at the transport and physical channels.

Transport Channels

Transport channels are of two types: dedicated and common. The channels listed in Table 7.2 are for the WCDMA_FDD mode of operation. There is only one dedicated channel and the remaining six are common. A dedicated channel (DCH) is one that supports the

Table 7.2   Transport channels in a WCDMA radio network

Channel	Abbreviation	Function/application
Broadcast channel (DL)	BCH	Transmits system- and cell-specific information
Common packet channel (UL)	CPCH	Transports packet-based user data in uplink direction
Dedicated channel (Ul/DL)	DCH	Transmits user or control information in either direction
Downlink shared channel (DL)	DSCH	Shared by several UEs; carries user and control information
Forward-access channel (DL)	FACH	Transports control information to the UE
Paging channel (DL)	PCH	Used to page information to the UE
Random-access channel (UL)	RACH	Received from complete cell and contains control information from the UE

soft-handover phenomenon discussed later in the chapter. Both dedicated and some common channels support fast power control.

Physical Channels

There are two types of physical channel: dedicated and common. The transport channels noted above are mapped on to the physical channels. These physical channels are a layered structure of radio frames and time slots carrying information related to the physical layers. The physical channels are identified by a specific carrier frequency, codes (channelisation/scrambling), timings, etc.

There are two dedicated channels and five common (see Table 7.3). The physical channels noted in this table are the ones on which the transport channels are mapped; there are in fact a few more common channels, such as synchronisation channels (SCH), common pilot channel (CPICH), paging indication channel (PICH), etc.

Table 7.3 Physical channels in a WCDMA radio network

Channel	Abbreviation	Function/application
Dedicated physical control channel (UL/DL)	DPCCH	Dedicated higher link information such as user data and signalling is carried on this layer
Dedicated physcial data channel (UL/DL)	DPDCH	Transmits dedicated physical-layer control information
Physcial random-access channel (UL)	PRACH	Transmits the data part of RACH and layer 1 control information
Physical common packet channel (UL)	PCPCH	Carries CPCH transport channel
Physical downlink shared channel (DL)	PDSCH	Carries DSCH transport channel
Primary common control physical channel (DL)	PCCPCH	Carries BCH transport channel and contains only data
Secondary common control physical channel (DL)	SCCPCH	Carries FACH and PCH transport channels

7.5.3 Uplink and Downlink Modulation

Modulation characteristics are different in the uplink and downlink directions. Both directions have a chip rate of 3.84 Mcps and use QPSK modulation. In the uplink direction, dedicated channels are multiplexed using complex coding schemes (also known as I/Q coding), while in the downlink direction they are multiplexed with respect to time. If time-based multiplexing is used in the uplink, then during the DTX, interference that occurs will be audible, which is not the case in the downlink because transmission is continuous with common channels.

7.5.4 Uplink and Downlink Spreading

Both uplink and downlink spreading are based on channelisation codes. The dedicated channel spreading factor variation is on a frame-by-frame basis in the uplink, while in the downlink it is not so except for the DSCH (downlink shared channel) that may use the variable spreading factor on a frame-by-frame basis. Uplink scrambling uses complex sequences of spreading codes. In the uplink there are two spreading codes, long and short, while in the downlink only long spreading codes are used. Long codes have a frame length of 10 ms, which means that a chip rate of 3.84 Mcps will result in 38 400 chips, whereas short codes have a length of 256 chips. As there are many spreading codes, a network planning tool should be used. Usually, only one scrambling code is used per sector in the downlink direction, as orthogonality between the channels (and users) needs to be maintained. Bad code planning will lead to less orthogonality and thus more interference, resulting in less coverage and capacity.

7.5.5 Code Planning

Owing to the spreading phenomenon, scrambling and channelisation code assignment becomes necessary. Uplink code allocation for both scrambling and channelisation is done by the system. As there are 64 code groups used in the downlink direction, having 8 codes each (i.e. a total of $64 \times 8 = 512$ codes), each cell that the user equipment has connectivity to should be assigned a different code. This brings a re-use factor of 64, which is quite high, making the cell search process less complicated. The high re-use factor also means that the code planning can be done manually.

7.5.6 Power Control

Fast power control tied with a high level of accuracy is an essential feature of WCDMA radio networks. As the frequency re-use factor is 1, so fast and accurate power control becomes even more essential. The reason is that, in the absence of a power control feature, the mobile station that is nearest to the base station can easily 'overshout' the other mobiles that are handled by the cell, thereby producing a blocking effect. Thus, a closed-loop power control feature is used.

In the uplink direction, the base station makes measurements of the power received from different mobiles in terms of SIR (signal-to-interference ratio). It then compares these with the target SIR. If $SIR_{measured}$ is greater than SIR_{target}, the BS will request the mobiles to reduce their transmitted power. A similar phenomenon happens in the downlink, but in this case the targets are the mobiles that are located on the cell edges, in order to provide them with sufficient power and reduce external cell interference.

Something else that is used in these networks is 'slow power control', also known as 'outer-loop power control'. This is used for controlling the SIR_{target} in a base station. The process is based on the needs of individual single links and is responsible for maintaining quality targets in the base station and the network. As the open outer-loop power control process is based on propagation aspects in the downlink, and is not sufficient as the uplink and downlink frequencies are quite different from each other, closed-loop power control is used. The BTS receives a signal from the UE and, based on its measurements, directs the mobile to increase or decrease its power level. And these measurements are based on the received signal level per bit to the interference signal level.

In the uplink direction, after the higher layers have set the uplink power, the UL control procedure starts. The power in the UL is controlled (or set) by the network. In fast power control, it adjusts the uplink power transmitted by the UE. This control is based on the SIR_{target}.

Another power control mode is the compressed mode. This is used in both the uplink and downlink directions, to provide faster power control. In this mode, larger steps are used so as to recover the power level for the SIR to move closer to SIR_{target}. (The compressed mode, or slotted mode, is needed to make inter-system or inter-frequency handovers. The mobile/BTS do not receive power control commands during compressed frames, and this is why larger steps need to be used. This is a special case that is used during the execution of inter-system measurements.)

Power control processes have a direct impact on the coverage and capacity of the network. Air-interface capacity in the downlink is more critical and is dependent on the transmitted interference. If the required transmitted power can be reduced to a bare minimum, the capacity can be increased. However, in the uplink, both the transmitted and received power will increase interference to, respectively, adjacent cells and users of the same cell. The better the isolation between the cells, the higher the capacity, and decreasing power levels to the minimum required can increase this isolation.

7.5.7 Handovers

WCDMA networks will experience a tremendous increase in the number of handovers taking place in their radio networks.

Handover is a phenomenon that takes place when the mobile subscriber is moving around. This is done so as to give the best coverage and quality to the subscriber. The following types of handover exists in WCDMA radio networks:

- intra-mode handover

- inter-mode handover.

Apart from this, there are two categories of handover: soft and hard. Handover was a phenomenon existing in GSM networks as well. In GSM networks, as every cell had a different frequency, the handover was called hard. In a WCDMA radio network, as the frequency re-use factor is 1 only, a different kind of handover is used, described as soft.

Intra-mode handover constitutes soft, softer, and hard handovers, whereas inter-mode handover will take place in a scenario where the FDD and TDD modes exist simultaneously. Here we will concentrate only on intra-mode handovers.

The intra-mode handover procedure is based on the measurements performed on the common pilot channel. When the mobile is in a position that is in between two cells (i.e. in a region of overlapping coverage), then it is connected simultaneously to two different sectors of two base stations (case 1 in Figure 7.6). A signal comes to the mobile from two

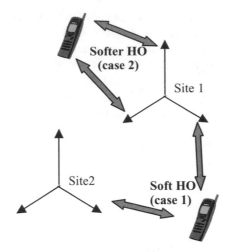

Figure 7.6 Handover in WCDMA radio network

directions, but with the use of RAKE receivers the mobile antenna receives a combined signal. Another type of handover is called 'softer'; in this case, the mobile is connected to two different sectors of the same base stations (case 2 in Figure 7.6). Again, the mobile receives both the signals owing to the presence of RAKE receivers.

In both the cases, as the mobile is using two different paths (interfaces) for communicating with the base station, two different codes are used so that the mobile can recognise the different paths. There is not much difference in the downlink direction, but is a significant difference in the uplink direction for the two handovers. In the case of soft handover, the received signal is routed to the RNC from both the base stations so that the best RNC can take a decision on keeping the signal which is giving a better quality. As this decision-making process takes time (about 10–80 ms), double air-interface capacity is used by a single mobile for one call. Hence, radio engineers have to take into account this process when capacity planning for the radio network. About 30% of the links witness soft handover, against 10% for softer handover. The measurements generally include received signal code power (signal power received on one code), signal power received within the channel bandwidth, and the ratio of the received bit energy to the thermal noise, E_c/N_0 (see below).

7.5.8 Coverage Planning

As mentioned above, coverage is an issue that is dependent upon the area to be covered, terrain type and the propagation conditions. Coverage should include all the regions such as dense urban, urban, rural, etc. As in GSM networks, all these factors have an impact on the distance that can be covered by a given cell (the cell range). But in contrast to GSM, capacity within a cell has a deeper impact on the coverage especially when the network is densely populated. A densely populated network would have more interference, having a negative impact on its quality.

Calculation of a cell's range can be done using link budget calculations. Link budget calculation in a GSM radio network is relatively simple. However, in a WCDMA radio network it becomes more complicated as the number of parameters affecting it directly increases. Let us look into these parameters before we look at an example of a link budget calculation. Some of the parameters are listed in Table 7.4. Most of these parameters are obvious, but some of them are new and have more impact in a WCDMA network.

Table 7.4   Equipment and network parameters

Equipment parameters	Network parameters
BTS receiver noise figure	Soft handover gain
BTS noise power	E_b/N_0
Cable loss + connector loss	Body loss
TX antenna gain	Processing gain
RX antenna gain	Interference margin
Thermal noise density	Power control headroom
Isotropic power	Peak EIRP

E_b/N_0 is the ratio of the received bit energy to the thermal noise.

E_b/N_0

E_b/N_0 is the ratio of the received bit energy to the thermal noise. E_b is received energy per bit multiplied by the bit rate. N_0 is the noise power density divided by bandwidth. Link budget calculations are basically done to calculate the E_b/N_0 ratio and the interference signal density.

Soft Handover Gain

The soft handover phenomenon gives an additional gain against the fast fading that takes place in the network. Due to the soft handover phenomenon, mobile connectivity is to a base station that gives a better signal quality. Thus due to macro diversity combinations, the soft handover gain has a positive impact on the base station.

Power Control Headroom

This is commonly known as the fast fading margin. It is the fade margin needed to maintain the closed-loop power control in action.

Loading Effect

Interference from neighbouring cells has an impact on the cell's performance. This is also known as cell loading, and the parameter to describe this is the *loading factor*. Owing to this loading, degradation in the link budget takes place, also known as *interference degradation*. If the loading factor is α, then the interference degradation margin can be calculated as:

$$L = 10 \log (1 - \alpha) \tag{7.1}$$

Theoretically, the parameter α may vary from zero to 100%, but practically it is in the range 40–50%.

The principle of the load factor is similar for uplink and downlink power budgets, but the parameters are different. In the downlink power budget, an *orthogonality factor* needs to be considered as well. As we saw at the beginning of this chapter, mobile subscribers are assigned codes and each of these codes is orthogonal to each other. Orthogonality of the codes is maintained only in direct propagation conditions, which is usually not experienced as there will always be some degree of multipath propagation. This will cause a delay in the signal and the mobile will treat some part of the incoming signal as an interference source. A perfect orthogonality would mean unity, while practically it may vary anywhere between 0.4 and 1.0.

Propagation Model
The propagation models described in Chapter 2, such as Okumara – Hata, Walfish – Ikegami, etc., are valid for WCDMA radio networks too. As in GSM networks, the correction factor procedure also needs to be followed.

Link Budget
An example of an uplink and downlink power budget is shown in Table 7.5. The fundamental principle remains the same as described in Chapter 2. However, in WCDMA radio networks the link budget calculations need to be done individually for voice and various data rates (e.g. 64 kbps, 144 kbps, 384 kbps) for a certain load (50% in this example) and for different applications.

7.5.9 Capacity Planning

Compared with GSM radio networks, capacity analysis finds itself more than ever dependent upon the coverage. The uplink and downlink coverages are quite different in a WCDMA radio network. The downlink coverage decreases with an increase in the number of mobile subscribers and their transmission rates. To achieve a service rate of 144 kbps, more sites are required than to achieve a service rate of 12.2 kbps. Moreover, as the number of users increases, the cell coverage area decreases. As downloading is more prevalent than uploading, the downlink capacity will be a critical factor for size of area covered by the cell. Once this has been dealt with, cell configurations can be decided upon in terms of sectors and carriers (TRXs).

Coverage and capacity plans are made and the process reiterated many times by using radio network planning tools (refer to Appendix A). Then the planning process moves forward to the site survey and site acquisition phase, as explained in Part 1. At this point it is helpful to understand a concept known as AMR (adaptive multi-rate), as an aid to choosing the right parameters to analyse when optimising the quality of the network.

7.5.10 Adaptive Multi-rate

Unlike in GSM, where speech codecs are fixed – e.g. FR (full rate) or HR (half rate) – and their channel protection is also fixed rate, in WCDMA radio networks the process of AMR is used to make it possible to adapt speech and channel coding rates according to the quality of the radio channel. This improves the error protection and channel quality.

The codec basically has one single integrated speech codec with eight source rates. This is controlled by the radio resource management functions of the RAN. These eight source rates are 12.2, 10.20, 7.95, 7.40, 6.70, 5.90, 5.15 and 4.75 kbps. The process of

Table 7.5 Link budget calculation in WCDMA radio network

Link Budgets:		Voice		LCD		UDD					
Data rate (kb/s):		12.2		64		64		144		384	
Load:		50%		50%		50%		50%		50%	
		Uplink	Downlink	Uplink	Downlink	Uplink	Downlink	Uplink	Downlink	Uplink	Downlink
RECEIVING END		Node B	UE	Node B	UE	Node B	UE	Node B	UE	Node B	UE
Thermal Noise Density	dBm/Hz	−174	−174	−174	−174	−174	−174	−174	−174	−174	−174
BTS Receiver Noise Figure	dB	3.00	8.00	3.00	8.00	3.00	8.00	3.00	8.00	3.00	8.00
BTS Receiver Noise Density	dBm/Hz	−171.00	−166.00	−171.00	−166.00	−171.00	−166.00	−171.00	−166.00	−171.00	−166.00
BTS Noise Power [NoW]	dBm	−105.16	−100.16	−105.16	−100.16	−105.16	−100.16	−105.16	−100.16	−105.16	−100.16
Required Eb/No	dB	4.00	6.50	2.00	5.50	2.00	5.50	1.50	5.00	1.00	4.50
Soft handover MDC gain	dB	0.00	1.20	0.00	1.20	0.00	1.20	0.00	1.20	0.00	1.20
Processing gain	dB	24.98	24.98	17.78	17.78	17.78	17.78	14.26	14.26	10.00	10.00
Interference margin (NR)	dB	3.01	3.01	3.01	3.01	3.01	3.01	3.01	3.01	3.01	3.01
Required BTS Ec/Io [q]	dB	−17.97	−16.67	−12.77	−10.47	−12.77	−10.47	−9.75	−7.45	−5.99	−3.69
Required Signal Power [S]	dBm	−123.13	−116.83	−117.93	−110.63	−117.93	−110.63	−114.91	−107.61	−111.15	−103.85
Cable loss	dB	2.50	2.50	2.50	2.50	2.50	2.50	2.50	2.50	2.50	2.50
Body loss	dB	0.00	5.00	0.00	0.00	0.00	0.00	0.00	0.00	0.00	0.00
Antenna gain RX	dBi	18.00	0.00	18.00	0.00	18.00	0.00	18.00	0.00	18.00	0.00
Soft handover gain	dB	2.00	2.00	2.00	2.00	2.00	2.00	2.00	2.00	2.00	2.00
Power control headroom	dB	3.00	0.00	3.00	0.00	3.00	0.00	3.00	0.00	3.00	0.00
Sensitivity	dBm	−137.63	−111.33	−132.43	−110.13	−132.43	−110.13	−129.41	−107.11	−125.65	−103.35
TRANSMITTING END		UE	Node B	UE	Node B	UE	Node B	UE	Node B	UE	Node B
Power per connection	dBm	21.00	27.30	21.00	28.30	21.00	28.30	26.00	33.30	26.00	33.30
Maximum Power per connection	dBm	21.00	40.00	21.00	40.00	21.00	40.00	26.00	40.00	26.00	40.00
Cable loss	dB	0.00	3.00	0.00	3.00	0.00	3.00	0.00	3.00	0.00	3.00
Body loss	dB	5.00	0.00	0.00	0.00	0.00	0.00	0.00	0.00	0.00	0.00
Antenna gain TX	dBi	0.00	18.00	0.00	18.00	0.00	18.00	0.00	18.00	0.00	18.00
Peak EIRP	dBm	16.00	42.30	21.00	43.30	21.00	43.30	26.00	48.30	26.00	48.30
Maximum Isotropic path loss	dB	153.63	166.33	153.43	165.13	153.43	165.13	155.41	162.11	151.65	158.35
Isotropic path loss to the cell border			153.63		153.43		153.43		155.41		151.65

AMR selection is based on the channel quality measurements for which both the UE and BTS are involved. Once it is confirmed that the quality of a signal is bad, the number of speech codec bits is reduced, thereby increasing the number of bits that can be used for error protection and correction. The speech frames in the AMR coder are of 20 ms, so if the sampling rate is 8000/s, 160 samples are processed. Thus, the bit rate can be changed at each of these 160 samples through in-band signalling or through DCH. This process is known as link adaptation (as seen in Chapter 6). Thus, by using AMR, speech, coverage and quality performance can be improved.

7.6 DETAILED PLANNING

When the pre-planning phase is over, and the search for sites has begun, the detailed planning process begins. This includes definition of the practical aspects of a site, such as antenna locations and heights, and study of link performances and interference analysis. Along with this, the all-important process of defining parameter settings takes place.

The detailed planning process is sometimes referred to as pre-launch optimisation, and radio network planning tools have an important role in this. Some data collected from drive tests is also used.

7.6.1 Coverage and Capacity

The foremost aspect of detailed planning is more accurate coverage and capacity plans. This starts with performance analysis of the radio links. The power budgets are analysed by taking some data from actual drive tests results. Factors that commonly cause a deviation of the practical link budget (i.e. the power actually received) from the theoretical calculations are:

- the propagation model

- link budget parameters

- propagation conditions.

All these factors are inter-dependent.

Fine-tuning of the propagation model is necessary under actual propagation conditions, as far as possible. A network covering a big region will require analysis for each part of the region separately, as propagation conditions (apart from topography) will change from one region to another. This leads to changes in the propagation model parameters, such as clutter type corrections, diffraction, topographical corrections, etc. One big change in a WCDMA radio network's link performance analysis is the fact that data play an important role. The types of data and their individual delay characteristics makes it impossible to define one set of parameters for the whole network. Each of the data types will have individual coverage probability requirements.

This analysis leads to results that give quite accurate antenna heights, antenna tilts, bearing, location of sites, etc. Both the uplink and downlink transmitted power need to be ascertained, after which the number of mobile subscribers can be more accurately predicted.

An accurate prediction of the number of subscribers will lead to a more accurate inter-ference analysis compared to the results in the pre-planning phase. The frequency re-use

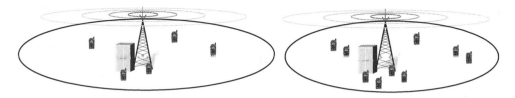

Figure 7.7 Cell breathing (decrease in coverage area with increase in subscribers)

factor is 1 in a WCDMA radio network, but code planning needs to done accurately. As there are 512 sets of codes, it is recommended that the automatic code allocation technique be used rather than manual allocation as the latter may lead to errors, thereby degrading the quality.

One concept that is important to understand at this point is *cell breathing*. The coverage area of a cell depends on the number of subscribers and their data rates. If the number of subscribers using the cell increases, the area covered by the cell decreases. As the load decreases, the area increases. This phenomenon is depicted in Figure 7.7.

Another phenomenon that directly affects capacity and quality is *soft capacity*. Erlang B tables can be used for the capacity calculation based on hardware limitations. However, in WCDMA radio networks, as the capacity is limited by interference (on the air interface), it is described as soft capacity. The less the interference from neighbouring cells, the more the capacity increases. The concept of soft capacity is important for data users. More capacity leads to a better connection (i.e. lower blocking) for real-time data users. Soft capacity can be obtained in one cell only if the adjacent cells have a lower loading (or fewer subscribers in them). Uplink soft capacity is based on the total interference at the base station, which includes own-cell and other-cell interference.

7.6.2 Radio Resource Management

For parameter planning in a WCDMA radio network, radio resource management (RRM) and mobility management (MM) should be understood. RRM and MM, as in GSM networks, consists of concepts such as the control of radio resources, admissions, power, handovers, etc. RRM is important from the perspective of air-interface resource utilisation, thereby offering optimum coverage and capacity and, above all, guaranteeing the quality of service.

As seen in Figure 7.3, the radio resource control (RRC) layer has two main layers beneath it which also constitute layer 2: RLC and MAC. MAC is an entity that forms the I_{ub} interface and channels the information coming in from the RLC layer. The RLC produces information related to RRC signalling, CS and PS data. The RRC has two main states: idle and connected. In the idle state, a subscriber is not connected to the network. In the connected state, user equipment is performing data transfer activity (voice, CS or PS). For the UE to change state from idle to connected, the call admission function comes into action.

The admission control (AC) resource management function controls the number of subscribers logged on to the network. It may admit or deny access to a new subscriber based on the resulting effect the extra subscriber would cause to the existing users. The result is based on algorithms that are executed in the RNC. These algorithms are also known as AC algorithms or CAC (call-admission control) algorithms. Admission control prevents congestion in the network and thereby maintains quality for the connected subscribers. AC algorithms are executed separately for uplink and downlink, but a requesting subscriber can be admitted only after gaining clearance from the uplink and downlink algorithms.

AC algorithms can be power-based or throughput-based. In the former, the uplink AC takes a decision based on whether the new subscriber mobile will increase the interference level over the planned load target and degrade the quality of the network. Each new user's entry affects the interference levels of the cell and adjacent cell. Thus, the interference level calculations are based on each of these cell measurements and the receiver noise. In throughout based calculations, both uplink and downlink AC take a decision on admission based on load factor calculations. The load factor has a direct effect on the interference margin (IM), as seen from the equation below:

$$\text{Interference margin} = 10 \log_{10}[1/(1 - L) \tag{7.2}$$

where L is the load factor from equation 7.1

Interference does not enter into these calculations, but does have an indirect effect. An enhanced functionality in these networks is congestion control. This feature monitors, detects and corrects situations when network congestion is degrading service quality. Overloading can be controlled by using functionalities such as fast power control (both uplink and downlink), handing over new subscribers (or the ones trying to congest the network) to another carrier, on even by reducing the throughput of packet data (a feature handled by packet scheduling). Moreover, mobiles that move within cell range cause different amounts of interference, so overload can occur when mobile(s) move towards the cell edge.

The power control function becomes all the more important in WCDMA radio networks operating in the FDD mode. The reason is that mobiles in WCDMA radio networks transmit continuously, unlike in GSM where every subscriber transmits in different time slots. Also, a WCDMA network uses only one frequency (the frequency re-use factor is 1). Each subscriber has an individual code and appears as a noise source to other subscribers. Hence, any inaccuracy in power control will lead to an increase in interference, which directly affects the number of subscribers getting admission to the network. This makes it necessary for power control to be accurate and fast. Thus, open-loop and closed-loop power control take place.

The power control feature is responsible for handover management in WCDMA radio networks. There are two main types of handover, soft (SHO) and hard (HHO). With SHO, a mobile subscriber is always connected to two base stations simultaneously, while with HHO the existing radio link is released before a new one is connected. The decision for handover lies within the RNC, where algorithms make decisions based on measurements received by the mobile stations.

Another important function in radio resource management is the allocation of codes. Owing to the use of a single frequency, separation of subscribers is based on these codes. The allocation of codes is based on information such as the network configuration and RRM features such as admission control (including also packet scheduling). The allocated codes have to be orthogonal to each other, but in practical conditions (with delays, etc.) the codes are not perfectly orthogonal, so interference may occur. Code allocation is done by the RNC in two steps: channelisation and spreading. The coding is done in such as way that the scrambling codes are different from each other, having low cross-correlation properties with OVSF channelisation codes being responsible for maintaining the orthogonality between the subscribers.

7.7 WCDMA RADIO NETWORK OPTIMISATION

The fundamental process of WCDMA radio network optimisation is quite similar to that for GSM. The process is, however, much more complicated and critical because of the presence of data, both real-time and non-real-time, with four classes, each class having its own types of application, with each application demanding a different quality aspect.

The process of optimisation begins in the early phase of network planning, right after pre-planning. It is slightly different from the network planning process, focusing more on performance optimisation by configuration changes (e.g. improving coverage by antenna height adjustments or by changing antenna tilts) than on launching.

The optimisation process starts with existing network data and the master plan made during the initial phase of network planning. Present network data are needed to ascertain whether there have been any deviations from the original plan during the implementation phase. The master plan will show the design targets (both short-term and long-term) for coverage, capacity and quality (Figure 7.8).

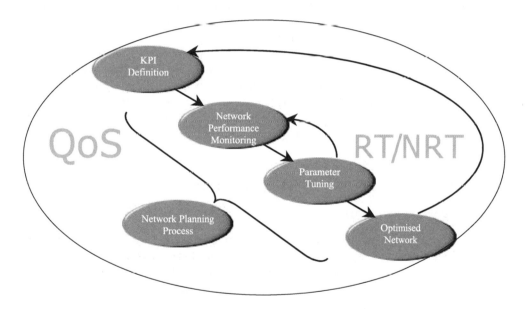

Figure 7.8 WCDMA radio network optimisation

7.7.1 Key Performance Indicators

Deciding on key performance indicators is the most important part of the process because it is here that the methodology of optimisation is decided. Owing to the complexity of a WCDMA network, there is an immense number of parameters. However, only a few of these parameters are chosen for optimisation, those that have the most significant impact on the radio network.

The initial values of parameters are usually derived from standard formulas. These key chosen parameters are the ones that are responsible directly for the coverage, capacity and quality of the network. The main KPIs may include call success/failure rate, dropped call rate, (soft) handover success rate, average throughput on uplink/downlink, and average throughput on various channels such as RACH, FACH and PCH, etc. When defining the KPIs, it is important to know whether or not tools are available to measure the performance of the chosen parameters.

7.7.2 Network Performance Monitoring

The network's performance can be monitored through drive tests and the network management system (NMS). In a WCDMA radio network, real-time data transfer delay is perhaps the most critical aspect to be monitored (see Table 7.6). Even within real-time data, each application has to be monitored/measured against the backdrop of its own QoS requirements.

With regard to the NMS measurements, as in GSM radio networks, most of the measurements are performed on the RNC, based on the key performance indicators on whose bases these measurements are run. The statistics collected from drive testing and the NMS are then used in conjunction with some post-processing tools that provide various types of output and reports.

The quality of the network is usually viewed from the perspective of mobile subscribers. This is why drive tests are important. The quality can be assessed right from the time when the first site goes live. The network management system usually comes into place when the network is near launch or has been launched, so can provide statistics when there are certain numbers of subscribers. The NMS should have the capability to handle the various aspects of the WCDMA radio network, which is in a multi-technology and multi-vendor environment.

7.7.3 Coverage, Capacity and Quality Enhancements

Coverage Enhancements

Coverage is directly related to link performance. An increase in coverage will demand an increase in the average TX power of a base station in the downlink direction. If the system capacity is downlink-limited, then an increase in coverage will lead to a decrease in capacity. If the system is uplink-limited, then the capacity is not affected. Thus, link performance increase is directly related to the increase in coverage.

There are many ways to improve the coverage. We have already seen the main parameters affecting link performance. Parameters such as block error rate, E_b/N_0, power control headroom, etc., directly affect the power budget and hence the coverage. Uplink coverage can be improved by decreasing the interference margin or by reducing the base station noise figure, or even by increasing the antenna gain. However, the processing gain and E_b/N_0 are the two major values affecting coverage.

So a reduction in E_b/N_0 increases the coverage of the network. This is because, for a lower E_b/N_0, less power is required for the same performance, so a bigger area can be covered. The performance of E_b/N_0 is dependent on a number of factors such as the bit rates, channel accuracy, SIR algorithms, etc. Uplink coverage becomes an issue at higher bit rates in WCDMA networks, so accurate traffic distribution will play a major role in coverage improvement. However, if the bit rate of the uplink direction can be reduced, the coverage

Table 7.6 Drive test result summary for Video Requirements

Date	Conditions					Required Performance					
	DPCH_Eb/ No (downlink) [dB]	DPCH_Eb/ No (uplink) [dB]	Loading	BLER (downlink)	Drop Call Rate[6]	Call setup failure rate[7]		Call set-up delay[8] 95th percentile[9]			
						Mobile originated call	Mobile terminated call	Mobile originated call	Mobile terminated call		
R1.5.2	>6.5	>4	<70%	<1%	<3.5%	<2.5%	<2.5%	<5 sec	<5 sec		
Drive test Results Summary				0.79%	1.78%	2.19%[10]	1.34%[11]	95.11%[12]	86.93%		
Sample Size				11997	354	549	571	634	505		

Call setup failure rate	Delay 95th perc.
2.19%	6.34 sec. (Alerting)

can be improved (as the transmitted power requirement would be less). This is possible only for NRT data (which are less delay-critical) and for voice signals, which have lower bit rates. However, E_b/N_0 cannot be lowered below the requirements of the requested service.

Another way to improve the E_b/N_0 ratio is to increase the multipath diversity. Two signals arriving at two antennas instead of one can be combined coherently, while the receiver noise can be combined non-coherently. This technique not only provides better gain, it also gives protection against fast fading etc. Concepts like antenna tilts are also used in WCDMA radio networks to improve the coverage area. A typical example is shown in Figure 7.9.

Site ID	Sector	Scrambling Code	Actual tilt angle		Proposed tilt angle	
			E tilt	M tilt	E tilt	M tilt
9205	A	172	6	0	4	0
9024	B	84	3	0	4	0
	C	92	5	0	6	0
9133	B	202	4	5	6	5

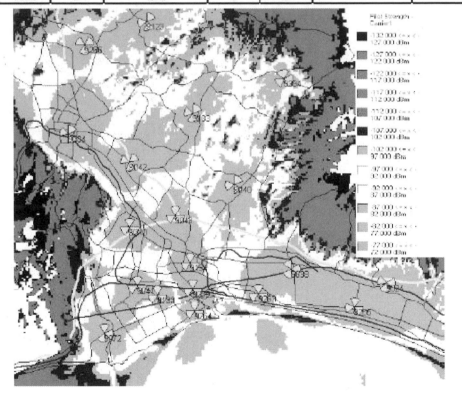

Figure 7.9 Typical coverage plans after antenna tilts in a WCDMA network

Capacity Enhancements

As noted earlier, capacity and coverage are heavily inter-dependent in WCDMA networks. The higher the uplink coverage, the lower is the uplink capacity, and vice versa. This is because lower capacity means fewer mobile subscribers, which means less interference. Moreover, the uplink power budget is used to calculate the cell range, which is further used to calculate the downlink power budget.

The load factor along with the link budget calculations can be used to study the capacity in the network. The load factor is used for capacity analysis for both the uplink and downlink directions. The load factor is dependent upon E_b/N_0, processing gain, interference, activity factor, etc. Orthogonality and soft handovers are further factors associated with the load factor in the downlink direction.

The best way to improve capacity is always to increase the number of cells/carriers. The increased number of sectors proportionately increases the capacity of the network. Additionally, orthogonal codes should ideally be truly orthogonal, but owing to multipaths some orthogonality is lost, thereby increasing interference. Multipath diversity improves the coverage but also reduces the orthogonality. Multipath diversity is more important at the cell edges as it improves the performance. Another way of improving capacity is by 'transmit diversity'. If multipath diversity is less, then the downlink transmit diversity increases the capacity to quite an extent. Lower bit rates would also increase the capacity. This is possible by using the adaptive mean rate (AMR) codes, AMR being the speech codec scheme that is used in UMTS.

Quality Enhancements

End-to-end QoS has to be considered in third-generation networks, but here we focus on the delay at the air-interface, which will have a direct impact on quality.

As noted in earlier chapters, QoS is application-dependent, but the immediate concern is to reduce the delay at the air interface for PS services. Unlike in GSM where voice quality is the only 'big' concern, in WCDMA attention turns to PS service requirements and performance. Delay may or may not take place at the air interface (it may be due to transmission or the core network), yet the first step for end-to-end (E2E) quality is the performance of the application at the air interface.

The example shown in Table 7.7 specifies the conditions and required performance for PS services. This is based on parameters such as BLER, DCR, application failure rate, delay and the throughput for 64, 128 and 384 kbps data rates. Drive tests results give the monitored results for these five parameters.

7.7.4 Parameter Tuning

As noted earlier, WCDMA networks have a huge number of parameters, and some of the key ones are chosen, measured/analysed and optimised. These parameters can be divided into various groups based on the functions they affect the most. These groups may include parameters affecting handover control, packet scheduling, power control, call admission control, etc.

Handover Optimisation

Soft handover gain is one of the parameters in the link budget calculations. SHO gives some protection against both slow and fast fading. With regard to slow fading, because of the

Table 7.7 PS service requirements and drive test results

Downlink Data rate [kb/s]	RAN	Conditions			Required Performance				
		Downlink DPCH_Eb/No [dB]	Uplink 64kb/s DPCH_Eb/No [dB]	Airlink capacity [kb/s per carrier/sector]	BLER (uplink and downlink)	Data session drop rate 1 drop/N hours	Data session activation failure rate	Session activation delay time[17]	Average Peak Throughput (uplink and downlink)[18]
64	R1.5.2	>5.0	>3.5	800	<10%	N>15 hours	<5%	<5s	>50 kb/s
128	R1.5.2	>5.0	>3.5	800	<10%	N>15 hours	<5%	<5s	>101 kb/s
384	R1.5.2	>5.0	>3.5	800	<10%	N>15 hours	<5%	<5s	>303 kb/s
Drive Test Results Summary					6.34%	No drop/ 15 hours	1.41%	89.40%	57.06kb/s (UL) 341.36kb/s (DL)
Sample Size					12320		594	538	14711(UL) 1194 (DL)

lack of correlation between base stations, a mobile is able to select a better base station (based on the measurements analysed in the RNC). With regard to fast fading, through the effect of macro diversity combining, the required E_b/N_0 is reduced. Soft handovers also induce overhead in capacity calculations, as at a given time a mobile is connected to more than one cell, thereby increasing the capacity requirements. Thus, both the overheads and the gain should be optimised.

The idea behind optimising the overheads is to save on downlink capacity. A typical value of the SHO overhead is 30–40%. SHO gain, on the other hand, can be estimated by using parameters such as DCR, CSR (call success rate), transmit and receive powers.

For capacity and coverage optimisation, it is very important to optimise the handover control feature in these networks. One important parameter to mention is the transmitting power of the CPICH. This parameter affects the coverage and it should be set as low as possible. The optimum value of this parameter will determine the coverage and capacity (i.e. should an extra user be permitted in the cell?). This parameter also affects packet scheduling. If the value of CPICH is not optimum, then either the network will be under-utilised or there will be huge interference if the number of users is more than planned, thus degrading the quality of the network. These parameters also affect the call success rate and dropped call rates. As in GSM networks, these two factors directly determine the network quality.

Packet Scheduling Optimisation

Packet scheduling is one of the most important aspects when controlling congestion in the network. Packet scheduling handles the non-real-time packet data, deciding on the timings of packet initiation and the rate at which packets should be delivered. As noted near the beginning of this chapter, four traffic classes that have been defined, and each of these classes has different applications to take care of.

The NRT packet data is bursty in nature, containing one or more data calls. Packet scheduling is done for both uplink and downlink for non-real-time bearers. Packets can be scheduled by using time-division or code-division techniques, or both. Packet scheduling and load control (inclusive of admission control) work in tandem. A higher load will lead to higher interference, which means fewer calls being admitted to the network. This affects the bits rate assigned to the NRT packet data. Thus, for packet scheduling (i.e. less delay and higher bit rates for NRT data), load control is an important parameter to analyse and optimise.

Transmitted power in the downlink and interference power in the uplink are further important parameters for optimisation. When the thresholds of these two parameters are crossed, preventive measures to control the load are initiated. From the perspective of packet scheduling this is important, because assigning higher or lower bit rates is dependent on the load control and AC see below).

Power Control Optimisation

Efficient and fast power control is the key to success of WCDMA technology. As noted earlier, power control is based on the SIR. Power control has a direct effect on the coverage area. Another aspect related to power control is interference that may lead to capacity limitation. Both uplink and downlink power control is necessary, with downlink control being the more critical. Power control of the common channels is necessary, the most important ones being the CPICH, AICH, PICH and CCPCH. However, this process is more one of control than of optimisation.

Admission Control Optimisation

The AC function is directly related to the load control process. The process is critical for both the RT and NRT traffic generators/users. As the AC function is power-based and throughput-based, parameters related to both those features are important for capacity and coverage optimisation. The two most important parameters are then the transmitted power and the received power, and orthogonality and throughput in both uplink and downlink directions.

8

3G Transmission Network Planning and Optimisation

8.1 BASICS OF TRANSMISSION NETWORK PLANNING

8.1.1 The Scope of Transmission Network Planning

Transmission network planning encompasses the interfaces between the BS and RNC, between the RNC and CN, and between RNCs, i.e. the I_{ub}, I_u (CS and PS) and I_{ur} interfaces respectively (see Figure 8.1).

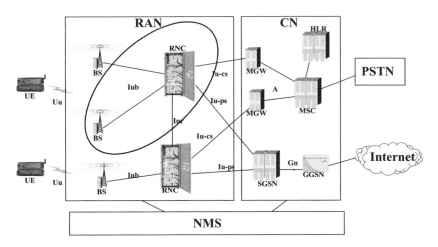

Figure 8.1 Scope of transmission network planning in a third-generation system (WCDMA)

Fundamentals of Cellular Network Planning & Optimisation A.R. Mishra.
© 2004 John Wiley & Sons, Ltd. ISBN: 0-470-86267-X

8.1.2   Elements in 3G Transmission Networks

Base Station (Node B)

Base stations play an important role in 3G transmission networks, and this role is different from GSM transmission networks. In the latter, if the number of tranceivers (TRXs) is known, the A_{bis} capacity calculation could be done. However, with data coming to play an important role, the components of the base station also change, thereby changing the dimensioning and functioning of the network.

A simplified block diagram is shown in Figure 8.2. A brief overview of the functions of these blocks is given below.

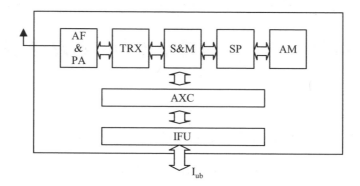

Figure 8.2 Block diagram of 3G base station

Antenna Filter and Power Amplifier (AF and PA)
The filter isolates the transmitted and received signals. The amplifier amplifies the received signals.

Transceivers
Transceivers contain one transmitter and two receivers. A TRX has two frequency bands that are separated by 190 MHz. The major differences from a GSM TRX include accurate power control, a different frame structure, and altered channelisation.

Summing and Multiplexing Units (S and M)
These sum the signals from the various other summing and multiplexing units and/or from signal processing units.

Signal Processor (SP)
The major function of this unit is to perform coding, decoding, and code channel processing (in both receiving and transmitting directions).

Application Manager (AM)
This unit is responsible for operation and management functions and carrier control. From the transmission network planning perspective, the AM takes the same place as a TRX in

GSM transmission network planning. The number of AMs is dependent on the traffic that is carried by the base station and is in no way dependent upon the number of cells or number of TRXs.

ATM Cross-connect (AXC)
This is responsible for cross-connection at the ATM (asynchronous transfer mode) level. It also acts as an interface between the AM and the interface units. AXC can be a standalone unit.

Interface Unit (IFU)
This is the interface between the base station and I_{ub} interface. It may vary depending on the type of transmission – E1/T1/JT1 or PDH/SDH, etc.

Radio Network Controller (RNC)

As noted in earlier chapters, the radio network controller in WCDMA 3G networks is equivalent to BSC of a radio network. Apart from similar functions, there is one main addition in RNC, which is an interface, I_{ur}, between two RNCs. The RNC is a very complicated network element, but a simplified block diagram is shown in Figure 8.3.

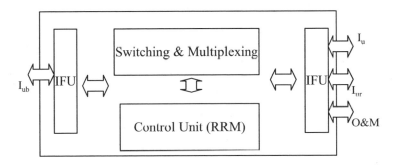

Figure 8.3 Block diagram of radio network controller (RNC)

Interface Unit (IFU)
There are mainly two kinds of interface unit, PDH and SDH. On either of the interfaces (I_{ub}/I_{ur}/I_u), a PDH or SDH interface can be configured.

Switching and Multiplexing Unit
Asynchronous transfer mode (ATM) forms the backbone of WCDMA networks. In the RNC, the functionalities of the switching unit include providing the required support for the ATM traffic, ATM cell switching, AAL2 switching, and multiplexing of traffic.

Control Unit (RRM)
Radio resource management is considered to be the most important function of the RNC. All three network planning entities – radio, transmission and core – are affected by it.

The control unit is responsible for RRM functions such as handovers, admission control, power control and load control. It is also responsible for the control mechanisms for packet scheduling and location-based services.

As cellular networks mature, location-based services (LBSs) are becoming more and more important, in both 2G and 3G networks (refer to Appendix C).

8.2 TRANSMISSION NETWORK PLANNING PROCESS

Fundamentally, the transmission planning process is the same as for GSM networks, but there is one important addition: the inclusion of ATM technology. The process again consists of nominal planning and detailed planning. Compared with GSM, dimensioning of a 3G transmission network is different, and in detailed planning there is ATM parameter setting to consider. All other areas are quite similar to those described for GSM transmission network planning. The process is shown in Figure 8.4.

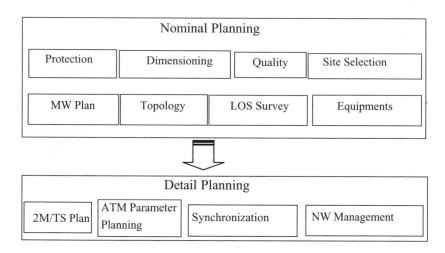

Figure 8.4 Transmission network planning process

NOTE: under 3GPP Rel'99, asynchronous transfer mode is the main technology in the transmission network. Later, IP is expected to replace ATM as the main technology.

A Note on Nominal Planning

Nominal planning is fundamentally the same as for GSM transmission planning – protection, link budget calculations, topology, equipment dimensioning, etc. The main difference is in dimensioning of the transmission network. Owing to the presence of voice, CS data and PS data traffic, with the data consisting of both RT and NRT traffic, dimensioning becomes much more complicated.

Before we venture into transmission network dimensioning, let us look at some factors that affect it.

8.3 ASYNCHRONOUS TRANSFER MODE (ATM)

B-ISDN (broadband integrated services digital network) provides digital connections capable of supporting rates greater than the primary rate (i.e. >2 Mbps) between user–network interfaces. B-ISDN was developed to support business and residential customers, with constant and variable data rates, data, voice and still and moving pictures, transmission and multimedia applications combining several service components. The idea behind the development of this technology was to provide a platform for development of all future technologies. Because of the variable nature of B-ISDN, ATM was selected as a switching and multiplexing technique as it is capable of giving the desired quality of service for different applications.

Information transfer takes place in small packets, also known as cells. One ATM cell is 53 bytes in length and is divided into two parts: header and payload. The header is 5 bytes long and the payload 48 bytes. These cells are multiplexed to form virtual channels, which are subsequently multiplexed to form virtual paths, as shown in Figure 8.5. Transmission media carry these virtual paths, which may vary from one to numerous (depending upon the capacity of the transmission media).

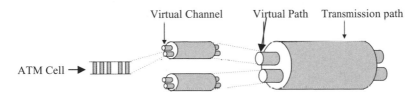

Figure 8.5 Multiplexing of ATM cell

8.3.1 Cell Structure

The cell structure of the ATM depends upon the type of interface. There are two types of interface. The one between the ATM terminal and ATM switch is known as the user-to-network interface (UNI), and that between two ATM switches is known as a network-to-network interface (NNI). Both types of cell have a 5-byte header and 48 bytes of payload, as shown in Figure 8.6. The difference lies in the structure of the header. The various components of header are described below.

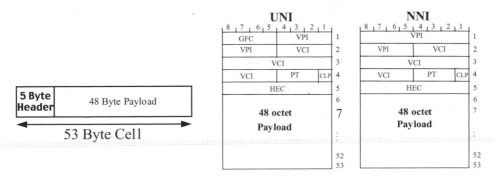

Figure 8.6 ATM cell structure

Header Error Control (HEC)

HEC is for error detection and 1-bit error correction of the 5-byte header. It is 8 bits in length and its sequence is calculated and set in the HEC field at the transmission side while the detected errors are corrected at the receiving side.

Identifiers for Virtual Path (VPI) and Virtual Channel (VCI)

As noted earlier, the ATM cell is multiplexed in virtual channels and virtual paths.

- A sequence of ATM cells belonging to a particular type of service or destination is a virtual channel.

- A virtual path is a number of virtual channels sharing one single link, which bundles different VCs into one VP. Also, to simplify routing and switching of cells belonging to a particular destination and particular type of service, VCs are combined into a single VP.

For the cell to be able to identify its virtual channel and virtual path, identifiers are needed. Cell routing is based on the values of VPI and VCI. The VPI has 8 bits for UNI and 12 bits for NNI.

Payload Type (PT)

The payload type basically indicates the type of information a cell is carrying (user information or management information). User information may be about congestion in the network, while management information may be to distinguish between different types of cells, such as whether it is an associated cell or a resource management cell.

Cell Loss Priority (CLP)

CLP is a 1-bit information field that gives the priority for cell loss, basically used for traffic management. There are two priorities for ATM cell loss: 0 and 1. Cells with CLP = 1 are discarded before cells having CLP = 0, in periods such as congestion.

Generic Flow Control (GFC)

This is a feature of a UNI cell header which basically helps the user to control the traffic flow for obtaining a desired quality of service. It is a 4-bit space in the UNI cell.

8.3.2 ATM Protocol Layers

ATM protocol layers (see Figure 8.7) are based on the B-ISDN protocol reference model consisting of three planes: user, control and management planes. The user plane handles user-related information such as flow control, while the control plane handles control-related information such as call control.

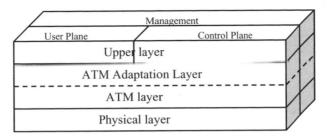

Figure 8.7 ATM protocol layer

Physical Layer

Generally it is said that the physical layer is the physical (transmission) medium. However, that is only a part of the physical layer. In the ATM protocol structure, the physical layer consists of two sub-layers: physical medium (PM) and transmission convergence (TC).

PM is the lowest sub-layer and its functions are dependent on the physical medium. The main functions usually include bit alignment, line coding and electrical/optical conversion. TC lies above the PM and its main functions include maintaining the cell rate of the ATM cells. It does this by adding/extracting cells to adapt the rate of ATM cells. Other functions of TC include cell delineation, frame generation and frame recovery.

ATM Layer

This is the layer in which virtual paths and virtual channels are established. It is responsible for providing the connection-oriented service (i.e. it establishes connections before transmission). The features of this layer are independent of the physical layer below it. It is responsible for generic flow control (only for the UNI) and generation (and extractions) of cell headers. Another important function includes cell VPI/VCI translations at the switching/cross-connect nodes. It is in this layer that cells from individual VPs and VCs are multiplexed into one resulting composite cell stream.

ATM Adaptation Layer (AAL)

As the name suggests, the ATM adaptation layer (AAL) 'adapts' the applications in a way to be understood by the ATM layers (or network) below. It also has two sub-layers: segmentation and reassembly (SAR) and convergence (CL). The SAR layer takes the cells from upper layers and sends them to the ATM layer. Another function of this layer is 'reassembly'. It reassembles the cells it receives from the ATM layer and passes the result to the CL layer. The convergence sub-layer is an interface between the application and the SAR. It splits the cells for transmission towards the SAR and, on receiving the information from the SAR, it reconstructs the cells into the original message. CL is a service-dependent sub-layer making AAL also a service-dependent layer. There are several AAL classes to support the different service classes. Connections falling under a service class have similar quality-of-service (QoS) requirements. The network can handle each service class separately by providing separate virtual paths. The following service classes are defined by the ATM forum:

- Constant bit rate (CBR): It required a static bandwidth that is available for the connection's lifetime. Traffic is fixed and synchronous. A typical example is real-time voice or video services.

- Real-time variable bit rate (RT-VBR): This is for services requiring a variable bit rate at the time of use. When the application is real-time, service falls under this category. A typical example is compressed video.

- Non-real-time variable bit rate (NRT-VBR): Another form to VBR, this is for applications that are not delay-critical. A typical example is e-mail.

- Available bit rate (ABR): Bursty traffic with a known bandwidth requirement falls under this category. Basically the traffic will follow whatever bandwidth the network can give to this. A typical example is Web browsing.

- Unspecified bit rate (UBR): Applications not having strict requirements for delay and delay variations fall under this category. It is also known as 'best-effort traffic', and there is no QoS guaranteed for this service class. Typical examples are e-mail and FTP.

Different types of AAL layers can handle different types of traffic. There are five AAL layers present, numbered from AAL1 to AAL5. They are described briefly below.

ATM Adaptation Layer 1 (AAL1)
This is used for traffic requiring a constant bit rate and a permanent connection during its execution (e.g. voice). The data should be transferred with minimum delay and high quality. Usually when data transfer is from one layer to another, some 'overheads' are added to the data stream. These overheads are basically some bits that are added to make the data stream 'layer' compatible. The larger the overhead (i.e. the higher the number of bits added), the longer is the delay. Overheads are kept to a minimum so as to reduce the delay.

ATM Adaptation Layer 2 (AAL2)
This is for variable bit-rate information that requires a strict relationship between the transmission and reception clocks. It provides bandwidth-efficient transmission for short, variable-length packets. It is able to multiplex the short packets from multiple users to one ATM connection. AAL2 is designed for applications such as compressed video (i.e. the RT-VBR type of traffic class). This layer can also do error-checking and sequencing of the data. Sometimes, when data are flowing (i.e. payload is empty), this layer also pads the payload so as to maintain the real-time aspect of the traffic.

ATM adaptation Layers 3 and 4 (AAL3/4)
Both these layers are used for data transfer. Another feature of these layers is multiplexing, i.e. allowing the same virtual channel to carry traffic from multiple sessions. This enables effective usage of the VC in applications such as X.25.

ATM Adaptation Layer 5 (AAL 5)
This is similar to layers 3 and 4 but more efficient. It can be used for both the message and stream mode of data transfer.

8.3.3 Multiplexing and Switching in the ATM

As ATM cells traverse the network, VCI and VPI values may change depending on the routing and addressing needs of the network and ATM elements. VCI and VPI have local significance only to each interface. The cells may enter a particular ATM switch with certain VPI/VCI values and leave with different values. VPI/VCI identify the next ATM pair of segments that the cells have to travel to reach the final destination. Depending on the level of the switching performance at these cells, the cells can be switched on either the VC or the VP level. If a cell is switched on the VC level, it is switched from one VP to another. If the switching is done on the VP level, all the VCs are included in a particular VP through the transmission path.

8.4 DIMENSIONING

We saw in Chapter 7 how the radio link budget changes with data rates. As the data rates increase, the coverage area decreases. Similarly, data rates have a big impact on transmission network planning right from the dimensioning phase. Once the number of base stations has been decided, all the RAN interfaces (except I_{ub}, I_{ur} and I_u) should be dimensioned, as should the number of RNCs required. For the dimensioning of interfaces, some knowledge of protocol stacks for CS data and PS data is required.

8.4.1 Protocol Stacks

Figure 8.8(a) is a protocol stack for the user-plane CS traffic. It basically shows the movement of the traffic from the mobile station to the MGW (media gateway). Figure 8.8(b) shows a similar movement of the traffic for PS data from a mobile station to the GGSN. Some hardware units in the equipment perform the functioning of these layers. One such example is node B (or base station). SP units perform the layer 1 functions while FP/AAL2/ATM layers are present in the application manager (as shown in Figure 8.2).

8.4.2 Overheads

One important point to note from the dimensioning perspective is that, when the traffic/signal moves from one layer to another, some overhead bits are added. These bits increase the capacity requirement of the interface. Thus, when doing dimensioning each overhead has to be carefully calculated.

Assume that a mobile subscriber is trying to establish a connection with the network. As the call is voice (CS), it will follow the protocols shown in Figure 8.8(a). The overheads to be attached to the call's capacity calculations would include voice activity factor (VAF), soft handover (SHO), protocol stack overhead, and signalling overhead.

- Voice activity factor (VAF) overhead: Once a mobile is connected to the network, the subscriber will either speak or listen (active or silent mode). In either case, some bits will be required for the functioning of that mode. Obviously, overheads during the active mode are more than in the silent mode. The type of voice connection can be defined in terms of AMR (adaptive multi-rate) codec modes. There are eight AMR

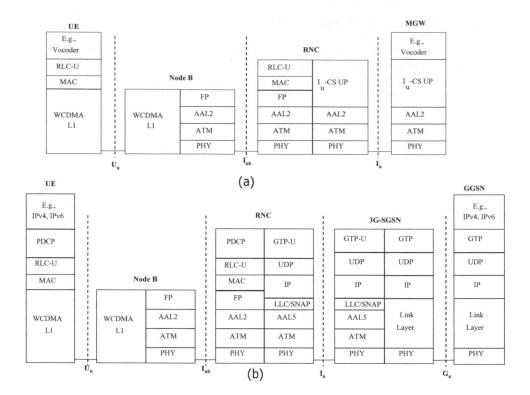

Figure 8.8 Protocol stack for (a) user-plane CS traffic, (b) user-plane PS traffic

codec rates, from AMR-0 to AMR-7, with the rates ranging from 12.2 kbps to 4.75 kbps respectively. Usually dimensioning for voice is done for AMR-0, or 12.2 kbps codec rate.

- Soft handover (SHO) overhead: As described in earlier chapters, soft handovers (which ensure continuous connectivity for the subscriber) take up more capacity. In some case the capacity requirement may go up to 150%.

- Protocol overhead: As the signal moves from one layer to another, some bits are added at each of these layers. This includes the RLC, FP and AAL2, etc., layers. The overhead will vary with the type of call (i.e. with the AMR codec).

- Signalling overhead: For each call, signalling is required. This signalling also needs capacity to associate itself with the data (CS/PS) when a call is being made from one subscriber to another. Hence, this capacity also needs to be added.

If a user is making a data (PS) call, then some of the above overheads will change depending on the type of call (i.e. the data rate). Each different type of data rate will have a different type of overhead. For a data call, however, VAF is not considered, as there will be no voice activity. However, overheads for re-transmission (explained in Chapter 6) and buffering are required. The amount of capacity required for buffering the user data is usually similar to that required for transmission.

Signalling on Transmission Interfaces

We have seen the general protocol structure of UTRAN in Chapter 7. Based on it, the following types of signalling are required at the various interfaces.

- ATM adaptation layer 2 signalling: AAL2 signalling is required for establishing, maintaining and terminating the AAL2 connections. It also performs functions such as error checking, or carrying signalling for lower ATM layers.

- NBAP (on I_{ub}): NBAP is of two types, common (C-NBAP) and dedicated (D-NBAP). C-NBAP is responsible for the handling of channels like RACH/FACH, cell configurations, fault management, etc. D-NBAP is required for the handling of dedicated/shared channels, air-interface fault management, configuration of radio links, etc.

- RNSAP (on I_{ur}): RNSAP (radio network sub-system application protocol) is responsible for functions such as radio link management, physical channel reconfiguration, measurements on dedicated resources, compressed mode control, power-drift correction, error indications, UL/DL signalling transfers, etc.

- RANAP (on I_u): RANAP (radio access network application protocol) is responsible for paging, flow control, processor/CCCH overload at the UTRAN, cipher mode control, location reporting, resource checking, direct transfer, trace invocation, error handling, handover procedures, etc.

Other Factors

When a call is transferred from the base station to the MSC (for voice) or towards the SGSN (for data), overheads are added. As the RRM functions are located in the RNC, there are no soft handovers at the I_{u-cs} interface, so no SHO factor is taken into account. However, factors such as blocking probability and signalling should be taken into account.

If the call is a data call, some additional overheads are added especially related to the GTP protocol apart from the overheads added due to UDP, IP, etc. (refer to Figure 8.8(b)) and those related to signalling.

RNC Dimensioning

Another focus area is RNC dimensioning. It is dependent on the capacity of the RNC to handle:

- interface traffic (e.g. $I_{ub}/I_{ur}/I_u$)

- the number of TRXs

- the number of BTS.

8.4.3 Example of Transmission Network Dimensioning

Consider an upcoming network for which transmission network dimensioning is to be done. The inputs required/given are as follows:

- *Dimensioning is to be done for phase 1 only.*

- *There are 10 000 subscribers each in the dense urban, urban, sub-urban and rural types of area.*

- *Each region has 50 base stations and each site has a configuration of 1 + 1 + 1.*

- *SHO = 40%.*

- *VAF = 67%.*

- *Total CS voice = 20 mErl per subscriber.*

- *Total CS data (64 kbps) = 1.50 mErl per subscriber.*

- *Total PS data (64 kbps) = 0.012 kbps per subscriber.*

- *Total PS data (128 kbps) = 0.272 kbps per subscriber.*

- *Total PS data (384 kbps) = 0.06 kbps per subscriber.*

Total I_{ub} traffic = CS (voice + data) traffic + PS traffic + signalling + overheads + O&M.

Assuming that the ATM overheads are 30%, signalling overheads in both the RAN and CN are 10% each, and the packet data overhead (GTP, IP, etc.) is 20%, the results are shown in Figure 8.9.

Area	subs	BS Type	TRX	Total Sites (BS)	Total TRX	CS Voice (kbps)	CS Data (kbps)	PS Data (kbps)	Total traffic/site (kbps)	Total Iub per site (kbps)	Total traffic to RNC (Mbps)
Phase 1											
D Urban	10000	1+1+1	3	50	150	178.3	349.4	0.0	527.70	1.2	58.0
Urban	10000	1+1+1	3	50	150	178.3	349.4	474.0	1001.70	1.2	58.0
Suburban	10000	1+1+1	3	50	150	178.3	349.4	474.0	1001.70	1.2	58.0
Rural	10000	1+1+1	3	50	150	178.3	349.4	474.0	1001.70	1.2	58.0
				200	600						

Area	No of Subs	BS Type	#Sites (BS)	CS Mbps	Iu to MSC CS Mbps RNC ->MSC	Iu to SGSN PS Mbps RNC -> SGSN	Iur Mbps RNC- RNC
Phase 1							
D Urban	10000	1+1+1	50	4.94	5.44	12.89	0.27
Urban	10000	1+1+1	50	4.94	5.44	12.89	0.27
Suburban	10000	1+1+1	50	4.94	5.44	12.89	0.27
Rural	10000	1+1+1	50	4.94	5.44	12.89	0.27
	40000		200		21.74	51.57	1.10

Figure 8.9 Example showing the results of transmission interface dimensioning

8.5 MICROWAVE LINK PLANNING

Microwave link planning is similar to that for GSM transmission planning. However, in 3G networks, ATM over microwave links seems to generate a concern over the quality of the links.

8.5.1 Error Rate and ATM Performance

In GSM transmission, a BER of 10^{-3} is considered to be good enough to meet the desired quality, but GSM transmission links carry more voice traffic than data. In 3G networks, with the traffic scenario changing (i.e. more data relative to voice), new standards are required for transmission systems. Moreover, networks in 3G are ATM-based. The main difference in microwave planning between GSM and 3G is the BER threshold consideration. As the quality requirement for data is more stringent, a BER threshold of 10^{-6} is considered during link planning. Standards such as ITU-T G.828, ITU-T I.356 an ITU-T I.357 are recommended to be used, as ATM requires high-quality transmission. The recommended values of threshold are 10^{-3} and 10^{-6} for PDH and SDH respectively.

Another aspect to be considered is the ATM performance itself. ATM performance is measured by using parameters such as 'cell loss ratio' (CLR). To keep this value to a minimum, availability targets are usually kept larger than 99.99%. Note, however, that this may depend on an operator's requirements.

8.5.2 Topology

Transmission networks in GSM saw many types of topologies: star, chain, loop, etc. In contrast, in 3G networks this will not be the case, at least in the beginning. There are two main reasons for this: capacity requirements and delay.

Traffic in 3G networks consists of voice, CS data, PS data and the common channels such as RACH, FACH, etc. If there are about seven users generating voice traffic, and three users generating CS-64 and PS-64 (both 64 kbps), then the total traffic generated by the base station (configuration 1 + 1 + 1) would be more than 50 Erl, of which common channels alone constitute more than 30 Erl, as shown in Figure 8.10. This increases the required capacity to almost four E1s, whereas in a GSM transmission network a configuration of 1 +

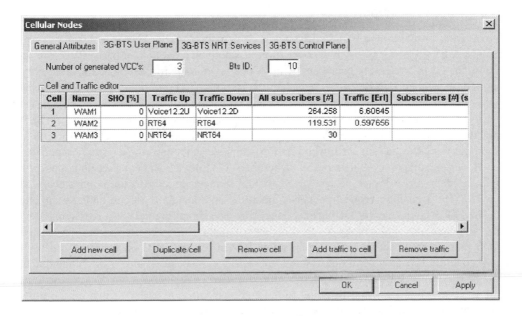

Figure 8.10 Example of traffic calculations

1 + 1 would need less than one E1 of capacity. Thus, longer chains/loops are not preferred in these networks.

Delay in another reason for making star the preferred topology in 3G transmission networks. For real-time traffic, especially data, delay will be the basis of quality. Longer chains mean more time required for the traffic to reach the RNC/SGSN, which will degrade the quality of the call. A star topology is used to reduce the delays.

8.6   DETAILED PLANNING

As shown in Figure 8.4, detailed planning has four major aspects: 2 Mb plans, parameter planning (ATM), synchronisation, and network management system planning. The principles of 2 Mb and NMS planning are the same as described in Chapter 3. Parameter planning is different.

8.6.1   Parameter Planning

Due to the involvement of ATM, parameter planning has now become an essential part of transmission network planning. Due to this, transmission planning in 3G networks has become a more complicated and challenging task as compared to transmission planning in GSM networks.

We have seen the basics of ATM in sections above. Now there are two main aspects to be focused upon in this part: ATM configuration parameters, and ATM performance (related) parameters. However, before we venture into parameter planning for the services provided by ATM network, let us understand the traffic management on the ATM itself.

8.6.2   Traffic Management on the ATM

Apart from virtual channels and virtual paths, as explained earlier, there are also PVCs and SVCs. A PVC (permanent virtual connection) is set up on a permanent basis. PVCs are set up by the NMS or element managers, while SVCs (switched virtual connections) require a signalling protocol (UNI/NNI) to set up the connection. ATM signalling is used to set up the virtual connections, allowing the switching and end station to communicate and exchange various management and QoS information for these connections. Signalling in ATM is always considered to be outband as there is always a dedicated pair of VCI/VPI values to send the signalling messages to an end station from a station requesting a virtual connection.

How does this process take place? When one ATM device wants to set up connection with another ATM device, it sends a signalling request directly to the connected ATM switch. The signalling request packets contain the end destination point address and any desired QoS parameter for the virtual connection. The signalling packet is reassembled by the switches and is examined to see whether it has the resources along with the desired QoS parameters to set up such a connection. It then sets up a VC on the input links and subsequently forwards the request out of the interface as specified in its switching table to the next ATM switch for further analysis, until it reaches the destination. 'V' connections are then set up along the path going to the end destination as every switch on the route examines the signalling packet and forwards it to the next switch. If a switch is not able to provide the desired QoS, the request is rejected and a message is then sent back to the point of origination. If the end point can support the desired QoS, it responds with accept messages as well as VCI/VPI values that the originator should use for the cells destined to the end point. Upon acceptance of the connections, traffic management functions ensure that each of the connections adheres to the traffic contract as agreed. This involves traffic management functions and techniques such as:

- traffic parameters

- connection admission control (CAC)

- conformance monitoring and control

- queuing.

Traffic Parameters

We have seen different types of service classes (CBR, UBR, etc.) available in ATM. Each of these service classes is defined by a set of traffic parameters, which describes the characteristic of the traffic source. These parameters are set to ensure a proper resource allocation to provide a guaranteed bandwidth and adhere to the desired QoS across the ATM network. The most common traffic parameters are as follows:

- Peak cell rate (PCR): This is defined as the maximum instantaneous rate that the user will transmit. It can be calculated as the inverse of the minimum time interval between cells. If the time interval between two cells is 1 μs, then the PCR is $1/(1 \times 10^{-6}\,\text{s}) = 10^6$ cells per second (cps).

- Sustainable cell rate (SCR): This is defined as the average rate of the cells, when measured over a period of time. It is also known as the mean cell rate.

- Maximum burst size (MBS): This is defined as the maximum number of cells that can be transmitted at the peak cell rate.

- Minimum cell rate (MCR): This is defined on the basis of the minimum cell rate that is required by the user. It is a rate that is negotiated by the end-system and the network such that the cell transmission rate never falls below the minimum specified/negotiated value.

- Cell-delay variation tolerance (CDVT): This is an error margin and defines the acceptable variation in the cell transmission time interval; i.e. it defines the upper bound in the variation in cell delay.

   Parameters like the above define the type of traffic and are thus called *traffic descriptors*. Before a connection is made, a *traffic contract* is negotiated based on factors such as traffic descriptors, CDVT, QoS class, conformance definition, etc. These parameters define the traffic type. However, the quality of the traffic is also an important aspect of ATM. The ATM forum has defined six parameters for this purpose. The first three listed below can be negotiated while last three cannot be negotiated.

- Cell loss ratio (CLR): This is the ratio of the number of cells lost (during transmission) to the total number of transmitted cells. The cells may be lost through error, congestion or even significant delays in their arrival.

- Cell transfer delay (CTD): When a cell is being transferred from the source to its destination, delay may occur due to propagation, queuing, etc. CTD can be defined as an average time taken by the cell to travel from source to destination inclusive of delays.

- Cell delay variance (CDV): Variation in the delay of cell transfer is measured by the CDV parameter. CDV can be defined as the difference between a measure of cell transfer delay and the mean cell transfer delay on the same connection.

- Cell error ratio (CER): This is the ratio of the cell(s) delivered with an error to the total number of cells.

- Severely errored cell block ratio (SECBR): This is the ratio of SES blocks (of cells) received to the total number of blocks transmitted.

- Cell mis-insertion ratio (CMR): Owing to errors in some headers, the cells may end up at the wrong destination. The CMR is the total number of mis-inserted cells observed during a specified time interval divided by the time interval duration (equivalently, the number of mis-inserted cells per connection second). Mis-inserted cells and time intervals associated with severely errored cell blocks are excluded from the calculation of cell mis-insertion rate.

ATM traffic parameters are related to the services offered by the ATM network. This is shown in Table 8.1.

Table 8.1 ATM service categories

Parameter traffic type	Traffic parameter	QoS parameter
CBR	PCR	CDV, CTD, CLR
RT-VBR	PCR, SCR, MBS	CDV, CTD, CLR
NRT-VBR	PCR, SCR, MBS	CLR
ABR	MCR	
UBR	PCR	

Connection Admission Control (CAC)

CAC algorithms determine whether a new connection is to be accepted or rejected (similar to AC in Chapter 7). A connection request is accepted only if sufficient resources are available and if connection will not affect the existing QoS. Some factors that are considered for a new connection request are:

- traffic parameters of new connections and QoS requirements

- existing traffic contracts and connections

- bandwidth, both allocated and unallocated.

Conformance Monitoring and Control

Conformance definition is responsible for indicating when a traffic contract is broken, e.g. if the network is not able to provide a bandwidth that is agreed, or a user exceeds the requested bit rate. This is done through two mechanisms: policing and traffic shaping.

Policing is a usage parameter function (UPF) which ensures that, during the connection, the network uses the traffic contract defined for the connection to check that it stays within the contracted service. If there are non-conforming cells, then the network takes appropriate action on them, such as setting the CLP bit of the non-conforming cells, thus making the cells eligible for discarding. This discarding is done to prevent any non-conforming cells affecting the QoS of conforming cells.

The traffic shaping function modifies the traffic flow and changes the characteristics of the user cell streams to achieve improved network efficiency and to get the lowest cell loss.

Queuing, Buffering, Cell Servicing and Congestion Control

To maintain the network's optimum performance, the ATM performs a series of cell treatment mechanisms; queuing, buffering, cell servicing and congestion control . Queuing will occur when two cells arrive at the same time and are going to the same destination. When cells of higher bit rate pass through a virtual connection with a lower bandwidth, congestion takes place. Buffering of cells occurs when two or more conforming cells are destined to the same output at the same time. Cells servicing, such as dropping of cells, occurs when a non-conforming cell with the CLP bit set to 1 arrives which causes congestion.

8.6.3 Network Element and Interface Configuration Parameters

Configuration of the network elements requires specification of a lot of parameters, which are related to the interfaces, ATM, IP, etc. Since ATM is used as a switching and multiplexing technique in 3G networks, every base station requires a unit that can handle ATM connections. This unit can be integrated with the base station (as shown in Figure 8.2) or can be a standalone unit. Parameter planning/setting is required for the interface units/hardware (PDH/SDH), ATM terminations/cross-connections, and those related to the IP addressing and synchronisation of the ATM cross-connect (AXC) unit.

Interface Unit Parameters

The interface unit parameters are related to the hardware and the PDH/SDH interface units that are used for transmission of ATM on the PDH or SDH. Hardware unit parameters basically are related to the type of system hardware that is being used, i.e. whether it is ETSI-based (E1) or ANSI-based (T1), etc. PDH/SDH interface parameters are related to the configuration of the PDH or SDH terminals that are being used. These also include some testing parameters apart from the ones needed to configure these terminals. As the structure of SDH is more complicated than PDH, more parameters are required for its configuration, related to the physical, multiplexer, regenerator and virtual container sections.

ATM Termination/Cross-connection Parameters

This constitutes the largest group of parameters to be configured. It includes parameters that are required for the identification and termination of virtual channels, virtual paths and their cross-connections. Also, parameters describing the type of traffic that is being carried by these virtual channels and paths are configured.

The 'rules' set out by the ATM forum or ITU recommendation for virtual channel and path identifiers should be followed, such as the number of VCIs reserved or the maximum/minimum number of bits that can be used for VCI/VPI. Every vendor/operator follows its own numbering scheme that should be respected.

Cross-connection in an ATM network can be at a physical level, VP level and VC level. (In GSM transmission networks, the cross-connections take place only at the physical level.) If expansion is expected soon, VP-level cross-connections should be planned as only few re-configurations are required even when new features are added. VC-level cross-connection are quite complex especially if only one VP is being used. One example is shown in Figure 8.11.

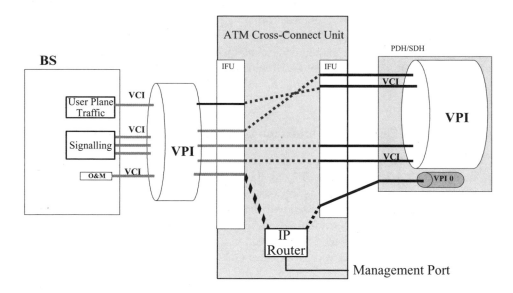

Figure 8.11 ATM terminations and cross-connections

IP Addressing Parameters

Both private and public IP addresses should be clearly defined. Private IP addresses are the ones that are used for internal communications and cannot be changed, while public IP addresses are those that are seen by other network elements and can be changed. Parameters related to the IP addresses and routing tables should be defined.

Synchronisation Parameters

Parameters related to the synchronisation of ATM cross-connections, such as definition of the clock source or synchronisation source, should be defined, along with the priority of these sources if there is more than one clock.

8.6.4 Summary of ATM Planning Features

A detailed understanding of ATM technology is necessary for ATM planning. ATM planning constitutes attention to the definition of:

- the types of traffic

- the VP and VC connections

- the number of VPs and VCs (i.e. VPIs and VCIs)

- VP and VC cross-connections

- physical, VP and VC connection parameters.

Some of these aspects are shown in Figure 8.12, in which 'grey' lines depict the physical paths, thick black lines show the VPCs, and thin black lines show the VCCs.

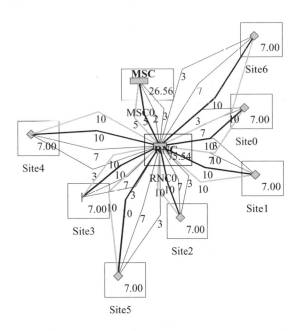

Figure 8.12 ATM planning

All these aspects of ATM planning are complicated, especially for a large network. In the network rollout phase, this process becomes complicated and tedious. For this reason, good ATM planning tools are used.

8.6.5 Synchronisation Plan

The design of the synchronisation plan for a 3G network is quite similar to that for a 2G network. The clock moves from the MSC to the BSC, from the BSC to the BTS, and from the BTS to the other network elements. The external clock (usually the primary reference clock, PRC) is applied to the MSC, which then distributes it further to the MGW, which sends it further to the RNC; and finally it is distributed to the base stations and other network elements. This is depicted in Figure 8.13.

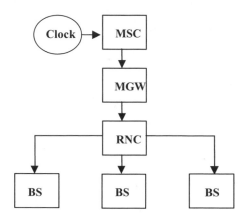

Figure 8.13 Synchronisation hierarchy

It is important to remember that, even in 3G networks, there might be some network elements that belong to a 2G network. In such cases, the ATM parameters need to be observed very carefully, as the 2G network elements are used to handling TDMA traffic and work with an assumption that traffic received is synchronous, which is not true for the ATM.

8.6.6 Network Management Plan

The last step in the design of the transmission network is the management plan. This will consist of management of communications of the DCN (data communication network) which will use the next hierarchal layer over the ATM, i.e. the IP layer. The plan will also have information about management of the transmission equipment, such as PDH radio and SDH equipment.

While doing DCN planning, the designer has to take into consideration many aspects, mainly the capacity of the I_{ub} interface. The capacity of the I_{ub} interface may range from 32 kbps to 128 kbps. Another aspect is the topology. The same topologies that are used in the network design can be used for DCN planning, including tree or chain topologies. The connections can be based on IP-over-ATM or pure IP. In either case, the IP addressing has to be such that there is always scope for future expansions, both in terms of network elements and technology. In a few years, fourth-generation technologies will have arrived, which are IP-based networks. Of course, equipment limitations have also to be taken into consideration when planning the IP addressing.

8.7 TRANSMISSION NETWORK OPTIMISATION

8.7.1 Basics of Transmission Network Optimisation

The transmission network optimisation process is fundamentally the same as for GSM transmission networks, but there are two additional considerations: parameter setting and quality of service. The process starts with defining the key performance indicators (KPIs) and the data collection process, followed by analysis of this collected data in terms of capacity, quality and parameter settings. The final optimisation plans are then made to meet the desired quality of service (see Figure 8.14).

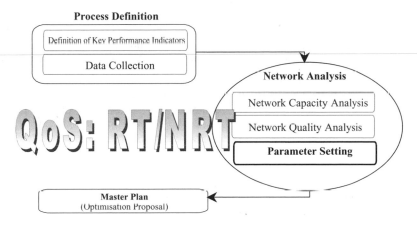

Figure 8.14 Optimisation process

8.7.2 Process Definition

Optimisation of a 3G network is aimed mainly at capacity and quality, against a backdrop of the QoS desired by the applications used by subscribers. In GSM there are very few key performance indicators, mainly related to the quality of the microwave network. However, as ATM is used in 3G transmission networks, the number of parameters increases tremendously, so increasing the number of KPIs. The main KPIs, apart from the microwave KPIs discussed in Chapter 3, relate to the ATM performance. ATM QoS parameters are the ones to be monitored mainly, and they act as KPIs for transmission network optimisation in 3G networks.

Data collection mainly involves existing network data, such as site configuration, network topology, link budget calculations, commissioning data, monitored data (i.e. power levels monitored at the time of site commissioning), and the performance of the KPIs. Most of the data collected is similar to the types we have previously seen in Chapter 3.

8.7.3 Network Analysis

This consists mainly of capacity analysis and quality analysis.

Capacity Analysis

For a GSM transmission network, the capacity calculation and assessment is relatively straightforward. If the site configuration of TRXs is known, the required capacity for the link can be calculated easily. However, capacity analysis of a 3G network it is not so direct. We have seen that capacity calculations are dependent on the types of traffic and so cannot be calculated directly from the number of TRXs or the number of application managers used in a base station. Thus, traffic parameters and the quality that is to be achieved are important inputs.

An increase in traffic is the most likely reason for analysis of capacity. Usually when the link capacity is dimensioned in the pre-planning phase, future needs are taken into account. In GSM transmission network optimisation, an increase in traffic would lead to an increase in the number of TRXs or the number of sites. In contrast, in 3G networks, capacity analysis is required more frequently because not only does traffic increase, the types of traffic (RT, NRT, etc.) change. An increase in traffic can lead to an increase in the number of signal processors/application managers and/or TRXs in a base station, and any increase in these will lead to an increase in link capacity.

We have already seen that a base station site with a small number of users generates substantial traffic. Even if the increased number of users is small, the packet data might increase (i.e. a small number of users generates PS traffic), and then both the number of AMs and link capacity will need to be increased. An increase in link capacity will mean that RNC capacity has to be re-examined.

Thus, capacity of the transmission network can be increased by:

- increasing the number of signal processors/application managers and/or TRXs

- increasing the number of base station sites

- increasing the number/capacity of RNCs.

Quality Optimisation

The asynchronous transfer mode was chosen for B-ISDN services so that quality of service could be guaranteed. Thus, ATM will be a focus area in 3G optimisation. Performance of the physical layer is another area for scrutiny.

8.7.4 Analysis of the ATM layer

Performance analysis of two ATM layers is of immediate interest: AAL2 and AAL5 (for CS and PS data respectively). The performance of these two layers can be monitored with the help of counters and performance indicators on management systems. These counters/measurements are mainly observed to calculate the error ratios and delays. In most of cases of degradation, delay is the reason. Factors that may cause degradation in quality due to delay are as follows:

- Propagation: Variation in propagation conditions may force the physical media to cause delays in transporting the bits in ATM cells between ATM nodes and switches. Another probable cause may be the performance of the physical media. We saw in Chapter 3 that propagation affects the performance of microwave links. Any degradation of the ESR and SESR will result in degradation in performance of the ATM layer as well.

- Traffic: ATM connections are designed for a given set of parameters. If the traffic load increases beyond what is expected, cell delays may result. In the light of this traffic increase, a higher QoS cannot be guaranteed.

- Architecture: Architecture of the network elements (e.g switches) can have a deep impact on the performance of the network. Factors such as buffer capacity, ATM cross-connection switching capacity and speed, etc., will affect the QoS.

It was noted earlier that there are three negotiable traffic (QoS) parameters in ATM: CLR, CTD and CDV. Optimum performance of these three parameters will lead of optimum performance of the ATM layer. The performance objectives to be met by the ATM layer are described in recommendation ITU-T I.356, which gives the provisional QoS class definitions and network performance objectives for parameters such as CDV, CLR, CMR, SECBR and CTD.

One important parameter that must be understood at this stage is CDVT. CDVT defines the upper bound in cell delay. This parameter is responsible for definition of the policing function in the network, that is, making sure that the traffic (or cells) reach their destinations within a required delay tolerance value. So basically, any cell that does not respect the service contract is discarded. For this purpose some algorithms are used. In general, CDVT can be calculated as the inverse of PCR. Transmission planning engineers should refer to recommendation ITU-T I.371 for more information on the appropriate values of CDVT.

Functions like *frame discard* and *partial packet discard* are used to avoid congestion and improve throughput. *Traffic shaping* is one of the mechanisms that will affect the cell stream characteristics to ensure a more predictable incoming traffic with lower cell loss. Thus, both efficiency and quality of the cell stream is maintained.

Performance of the physical layer is another key issue. There are no objectives specified for SES_{ATM} or ATM availability (except in the definitions). SES_{ATM} is defined as a second when $CLR > 1/1024 = 9.8 \times 10^{-4}$ or $SECBR > 1/32 = 3.1 \times 10^{-2}$ as from ITU-T Recommendation I.357. Based on simulations, the equivalent worst-case BER for SES_{ATM} is 3.4×10^{-5} and the limiting parameter is $SECBR$.

Performances and unavailability of ATM connections are related to this threshold value. Some examples are shown in Tables 8.2 and 8.3.

Table 8.2   Access network performance objectives in ITU-T G.826

ITU-R Rec. F.1189-1: *Performance Objectives for a National Access System*; **from G.828**								
Radio capacity	Nearest SDH container	*ESR*	*SESR*	*SES*/month	*BBER*	*BER*$_{ESR}$	*BER*$_{SES}$	*BER*$_{BBER}$
PDH 2E1	VC-12	0.0034	0.00017	442	1.7E-05	1E-09	1.7E-04	2E-08
PDH 4E1	VC-2	0.0043	0.00017	442	1.7E-05	7E-10	1E-04	5E-09
PDH 8E1/16E1	VC-3	0.0064	0.00017	442	1.7E-05	8E-10	4E-05	3E-09
STM-1	VC-4	0.0136	0.00017	442	1.7E-05	1E-10	2E-05	6E-10

Fixed block allocation $C = 8.50\%$
Access length L, rounded to nearest 500 km $= 500$ km

Table 8.3   Access network performances objectives as from ITU-T G.826

From I.356

Parameter	Allowance	HRX 27 500 km	Access network	Scaling ref.	Equivalent *BER*
CLR	24%	3.00E-07	7.20E-08	10.67	1.70E-11
CER	9.50%	4.00E-06	3.80E-07	383.40	9.91E-10
SECBR	9.50%	1.00E-04	9.50E-06		
SES$_{ATM}$	–	–	–	–	3.40E-05

8.7.5   Parameter Setting

The inputs of the whole analysis process can be divided into three major categories. Dimensioning parameters form the first inputs. The second inputs come from the actual implemented network, such as the capacity of the links and the media used, along with the synchronisation and network management settings. The third inputs are in the form of ATM parameters and radio network parameters (see Figure 8.15).

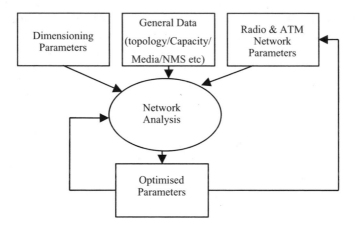

Figure 8.15 Process to study the impact of radio parameters on the transmission network

The dimensioning parameters play a very important part in the whole process. The network is initially dimensioned based on factors such as existing knowledge and equipment limitations at the time when the actual network is launched. Then lots of factors and parameters change, so it becomes necessary to make a reassessment of the dimensioning.

The general data consist of the actual implemented and commissioned data, such as topology, link budget calculations, interference analysis, capacities of the links, etc., along with the actual received power, system curves and recoded fade margins. Also, for long hops, fade margins can be recorded over a certain period. Availability performance data can also be recorded for a longer period.

The radio network parameters are again subdivided into dimensioned and optimised parameters. Any effects of the optimised radio parameters can be seen on the transmission network parameters. In this way, later on down the process, the relationship between certain radio parameters and the transmission network performance can be examined, allowing the setting of these parameters such that the performances of *both* the radio and transmission networks are optimal for best E2E (end-to-end) QoS for the network. Once this has been done, the optimised parameters are looped back again to the network analysis so as to cross-check whether they are giving the desired network performance.

9

3G Core Network Planning and Optimisation

9.1 BASICS OF CORE NETWORK PLANNING

Core network planning of a third-generation (WCDMA) system consists of both circuit core and packet core planning.

9.1.1 The Scope of Core Network Planning

The scope of core network planning includes dimensioning of network elements such as MGW, MSC (and VLR) and HLR (and AC/EIR) from the CS side, while the PS core will include SGSN, GGSN along with interfaces.

9.1.2 Elements in the Core Network

Network elements in the 3G core network remain almost the same as seen in earlier chapters on core networks, except that two new concepts are added: MSC and MGW (media gateway). These are briefly explained below (see Figure 9.1).

According to 3GPP Rel'4, the control and user-plane traffic is separated, with MSC and MGW handling the control plane and user-plane traffic respectively.

The MSC (also called an MSC server in 3GPP Rel'4) handles the call and mobility control function for CS traffic generated by registered subscribers, incoming calls from other networks, mobile-originated calls, and mobile-terminated calls. Apart from this, it is also responsible for signalling conversion (user-to-network signalling converted to network-to-network signalling). The MSC is also responsible for controlling the MGW. The MSC

Fundamentals of Cellular Network Planning & Optimisation A.R. Mishra.
© 2004 John Wiley & Sons, Ltd. ISBN: 0-470-86267-X

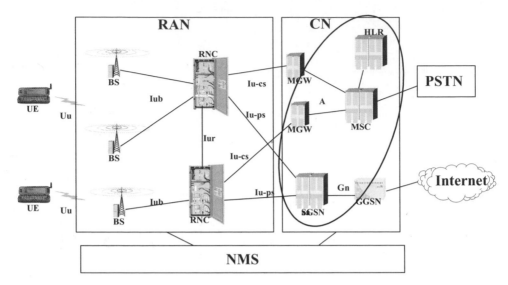

Figure 9.1 Scope of core network planning in a third-generation system (WCDMA)

server, as in GSM, contains the VLR as well as other elements such as a group switch (GSW), etc.

The MGW is responsible for user-plane traffic handling. The main function is switching voice and data towards the required destinations, converting the ATM traffic into time-division mode and vice versa. It is able to handle/switch both the switch and data traffic. I_{u-cs} signalling termination also takes place on the MGW. Both the versions of IP (4.0 and 6.0) are supported by the MGW.

9.2 CORE NETWORK PLANNING PROCESS

The processes of network planning for both the CS core and PS core essentially remain the same as described in earlier chapters (see Figure 9.2). However, since there are changes in the radio network elements, transmission network elements and interfaces, core network planning is affected. Network elements, interfaces and signalling plans are the main areas of focus in the core network plan. In the following sections, we will go through the process and changes in the CS and PS cores in turn.

9.2.1 Circuit Switch – Core Network Planning

Although CS core network planning involves both switch and signalling network planning (as in GSM), in a WCDMA 3G network there are certain differences related mainly to the changes mentioned above (i.e. new network elements and interfaces). The process begins with network analysis and dimensioning, followed by detailed planning to produce the final CS core network plan.

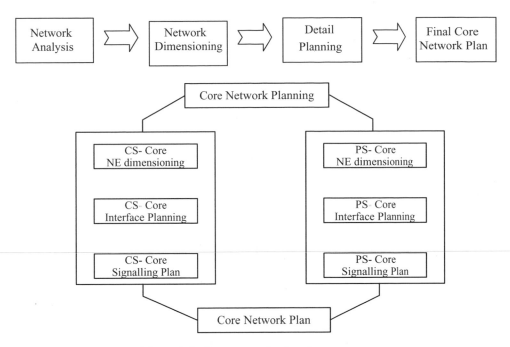

Figure 9.2 Core network planning process

Network Analysis

Information collection, again, an important part of the process. Information to be collected includes the existing/designed network topology (both radio and transmission), subscriber database (present and forecasted), topography of the region, traffic measurements etc. Quality and performance targets give direction to the analysis and design of the core network. In GSM core network planning, the MSC (switch) was one important network element around which the whole design process revolved. In 3G, as the functionalities extend to MGW, the whole process will revolve around the two network elements, MSC and MGW, their inter-connection and the planning of the signalling network.

Network Dimensioning

The existing subscriber base and expected growth rate are important factors. Dimensioning should be done in such a way that, when implemented, the network will be able to handle the expanding traffic for the next few years. Network planning tools are used (see Appendix A).

In third-generation networks, a twofold approach is taken for dimensioning the CS core network, that is, the network is dimensioned with respect to both the user plane and the control plane. And as we have seen, the MGW and MSC (or MSC server) are two elements that are responsible for the user plane and control plane traffic. Thus, dimensioning of the CS core involves dimensioning of the number of MGWs, the number of MSCs, and the signalling related to them.

Dimensioning the number of MGWs is dependent on several factors, which mainly include the traffic distributions, subscriber database, location of elements like the MSC,

RNC and external (other network) elements. The number of MGWs can also be calculated by the amount of area that needs to be covered by a given MGW. The locations of the MGWs is another factor. An MGW can be distributed in the network (near the RNCs) or can be co-located near the core site. A distributed structure will result in transmission saving, but on the other hand it also leads to an increase in the number of MGWs, which in turn results in more inter-MGW connections. This increases the complexity of the network, making it difficult to maintain or expand. A concentrated structure decreases the equipment costs but increases the transmission costs. (Readers can draw the analogy from Chapter 3, where the concept of BSC location is discussed.) Also, in a concentrated structure, the complexity is less. In network rollout plans, both types of structure, distruted and concentrated, might be used simultaneously depending upon the requirements.

Dimensioning of the number of MSCs (or MSC servers) depends directly on factors such as the subscriber base, their calling behaviour and the number of RNCs. The reason is that each MSC contains VLR in a limited capacity (i.e. can handle a specified number of subscribers and the number of attempts to log into the network). Every RNC can be connected to only one MSC.

Once the subscriber base is known for the network, the traffic that is expected to flow in the CS network is calculated, using an approach similar to that outlined in Chapter 4. The traffic matrix will contain the traffic flowing in the network due to calls generated internally, and expected incoming calls generated in other networks. One small example is shown in Figure 9.3. Unlike a similar figure shown earlier (Figure 4.3), the traffic matrix contains MGW as the network element instead of MSC.

From	MGW	External Network
MGW	X	800
External Network	1600	X

Figure 9.3 Traffic generation (in Erlangs)

9.2.2 Packet Switch – Core Network Planning

Network Analysis

The packet core network is similar to that in GPRS network planning (refer to Chapter 5). The main elements include SGSN, GGSN, border gateway (BG), domain name servers (DNS), LIG, switches and routers, charging gateways (CG), firewalls (FW), and the G-interfaces as shown in Figure 9.4. (Note: Fig. 9.4 forms part of Fig. 5.1).

Network Dimensioning

We have already seen most of these elements in Chapter 4, but there are some differences in 2.5G and 3G packet core networks. One is obviously the data rate, which is higher in 3G. The evolution of networks also brings new challenges with the increase of the number of PDP

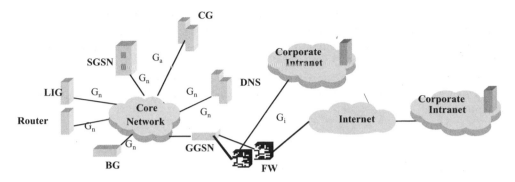

Figure 9.4 Elements and interfaces to focus on in 3G packet core network planning

contexts. In early releases of GPRS, there is only one PDP context per access point for one subscriber. However, with the new releases and quality-of-service (QoS) differentiation, the network is able to handle several PDP contexts per access point for one subscriber. This has an impact on the GSN dimensioning as discussed in Chapter 5 (in transmission network dimensioning). As data increases in 3G in terms of type, rates, etc., the SGSN and GGSN dimensioning becomes more complex ******* them in previous chapters. Predictions for the number of subscribers, voice and data, are very important, as they give an indication on the PDP context and data traffic that will be needed in the dimensioning process.

Although overheads were taken into account in 2G networks, in 3G networks they play a more crucial role. In fact, the impact of overheads in quite substantial in all three parts of planning, for radio, transmission and core. In the packet core network, overheads affect the capacity and subsequently the quality of the network. Overheads in the packet core network can be calculated on the basis of the protocol stack shown in Figure 9.5. The overheads due to each layer needs to added to during the dimensioning phase. Layer 1 (L1) is the physical interface while layer 2 (L2) is the ATM (or Ethernet) layer. The percentages of overheads due to each of these layers depend upon the traffic.

Dimensioning of border gateways (BG) and firewalls (FW) is another part of core network planning. Any given 3G network will have diverse types of data traffic, and will interact with different types in other networks. Thus, firewalls become more important in 3G networks.

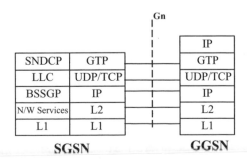

Figure 9.5 Protocol stack in a PS network

BG dimensioning is dependent on the number of 'own' subscribers in external networks and external subscribers in the 'own' network. Firewall dimensioning may not be a part of the initial network plan, but it will be required in the later stages to protect the network from 'intruders'.

Core network planning engineers should remember that other factors – such as redundancy of equipment – is highly recommended in the core network, but cost implications should be considered as well.

9.3 DETAILED NETWORK PLANNING

9.3.1 Circuit Switch (CS) Core Network

The detailed master plan for the CS core will include plans related to routing, signalling and parameter settings. All these are based on the core network plan that is an output of network dimensioning (though it may change in the detailed planning phase).

Elements

The network plan consists of the CS core network elements such as MSC, MGW and the inter-connections of these elements amongst themselves and with other elements in the cellular network. A CS voice network plan, as shown in Figure 9.6, contains the connection between the MGW and MSC. This plan also indicates the capacity required for the link (four E1s in the example). Apart from this, the routing plan for both voice and signalling is required. This plan contains primary routes and secondary routes for carrying the voice and signalling traffic. A load sharing factor is also decided.

The locations of the MGWs/MSCs, inter-connections between them, connection between the 'own' network and external networks, etc., should be decided. In the core network, elements can be inter-connected physically or logically or in both ways. In the present scenario of 3G core networks, physical connections can be made using LANs, IP trunks, etc. depending on the type of connection switches (or facility) the core network element has. Aspects such as planning of IP addressing schemes usually accompany this.

The naming and numbering conventions remain similar to those in GSM core network planning, with destinations, sub-destinations and circuit groups clearly defined. Signalling points, signalling links, signalling link sets, etc., are defined also in the detailed plan. Signalling plans are based on the concepts discussed in Chapter 4. Differentiation between the subscribers of different networks can be done by IMSI analysis.

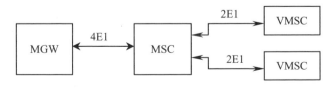

Figure 9.6 Simplified example of CS voice network plan

Example 1: Naming of the circuit groups

Assume that a network shown in example is coming up for an operator Cite-Telecom.

Circuit groups can be named as: CTTMSC2

Where CT is the name of the operator (CT); MSC gives the destination code, while 2 stands for the exchange terminal.

Based on the scheme decided by the core network engineers and network operator taking into consideration equipment specifications.

Transit Layer

Planning of the transit layer (a concept similar to that of transit layer planning in Chapter 4) is another concept to be taken care of during the detailed planning process. We have seen earlier the necessity for a transit layer; however, in a third-generation CS core network the transit layer is planned for MGWs and MSCs instead of just MSCs in GSM core networks. The planning fundamentals remain the same, but note that the need for a transit layer applies only in the ATM-backbone network, not in an IP-backbone network. One such example, with the traffic, is shown in Figure 9.7.

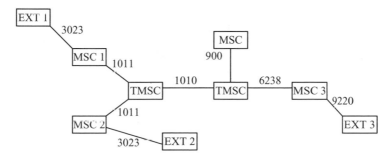

Figure 9.7 Traffic routing

Parameter Planning

The final aspect of detailed planning is parameter planning. Parameters are defined for configuration of the exchange terminals, traffic routing, signalling, circuit groups, destination and sub-destinations, etc. Although the number of parameters defined is substantial (depending on equipment specifications, etc.), some commonly used parameters are noted below.

Configuration of Exchange Terminals
Parameters in this regard include the definition of ports for termination of the PCMs, and those related to synchronisation for the core network elements. Limitations of the equipments and naming conventions should be respected when giving the numbers for it. A synchronisation hierarchy should be clearly defined.

Routing Parameters

Routing parameters give the direction of traffic from the switch. Other parameters are related to outgoing calls for their signalling registration, control parameters, and the starting point for a call.

Destination and Sub-destination Parameters

Parameters related to the call's destination, and the route it takes, are required. The call may be routed through one or more sub-destinations which need to be defined in the naming conventions. Also, the parameters related to load sharing have to be clearly defined. Based on these parameters, the call will take the shortest route to reach the desired destination. The parameters generally include the name destinations (sub-destinations), route identifiers, types of calls, etc.

Circuit Group Parameters

Circuit group names, types and identifiers should be defined in the parameter lists. Also, parameters associated with the functioning of the circuit groups such as the signalling registers, SPCs, control, destination, etc., are defined.

Call Parameters

Parameters related to calls need to be defined. These include:

- network-related parameters such as country code (e.g. 34 for Spain), and codes required for making national and international calls

- definitions of IMSI (such as indicators, range) and associated public land mobile network (PLMN)

- call identity parameters such as calling and called party status (e.g. called subscriber is roaming or in conversation mode, etc.)

- network- and switch-related parameters like mobile country code and the mobile national code for the operator (e.g. 12 for an particular operator), etc.

Parameters required for call charging include call type (e.g. if the subscriber is using the network for a roaming call or for sending an SMS), whether it is a free or chargeable call, whether the incoming call is from an own-network subscriber or from an external-network subscriber, etc.

Signalling in the CS Core Network

RANAP signalling exists between the RNC and the MGW (i.e. on the I_{u-cs} interface) and is carried to the MSC. The signalling protocol on the I_u interface is shown in Figure 9.8. RANAP messages are embedded in MTP3SL messages. SS7 signalling network services over the ATM network require signalling AAL (SAAL) to deliver MTP3 signalling messages between network elements. SAAL is responsible for the correct transfer of signalling messages on an ATM-based SS7 signalling link. Each of these sub-layers is considered when doing the traffic calculations and parameter settings.

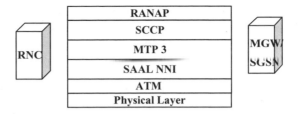

Figure 9.8 SS7 signalling on the I_u interface

The signalling load calculation forms an important part of CS core network planning. The factors affecting the RANAP signalling load are the amount and type of traffic generated by the subscribers, and the number of subscribers. Based on these, the bandwidth required for signalling is calculated. When calculating the signalling load towards the MSC (from the MGW; also known as 'SIGTRAN'), overheads must be included. The signalling parameter group consists of signalling points, signalling point codes, signalling links, signalling link sets, etc. Also included are the signalling routing definitions.

9.3.2 Packet Switch (PS) Core Network

As shown in Figure 9.4, the core network is quite similar to the one discussed in Chapter 5. Here we will look at two aspects of detailed PS core network planning, namely:

- connecting the (PS) core network elements to each other (i.e. planning for the interfaces)

- IP addressing and traffic routing.

As with any core network, equipment specifications play an important role not only in dimensioning but also in detailed planning. In a packet core network, SGSN and GGSN capabilities are the most important defining factors, so it is important to understand the functioning of these two network elements.

As shown in Figure 9.9, SGSN contains four main blocks: interface units (IF), router, GPLC/GTPU and management. An interface unit is required for handling the data

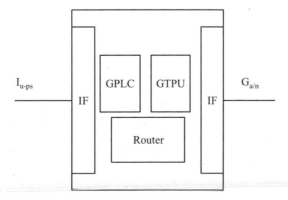

Figure 9.9 Simplified block diagram of SGSN

connections in the direction of the RNC on one side and the GSN (SGSN/GGSN) on the other side. A GPLC card is also present and is used for receiving, sending and managing the packet data. The router in the SGSN performs the functions of both a router and a processor. In fact, this is the unit that 'controls' the SGSN. All the units in SGSN are usually Ethernet or ATM interfaces and are connected to the router. The traffic movement is guided by IP addresses. GGSN is much simpler than SGSN as it can be termed as a simple router. SGSN is also responsible for the mobility management-related functions. The functionality of the GGSN router is only routing without much processing to be done, unlike in SGSN where the router performs processing functions as well. The main function of the GGSN is to generate the PDP contexts and to set up the GTP tunnel. Once this is done, all the GGSN has to do is to maintain them. Again, IP addresses play an important part in the process, as they are required for generation of the PDP contexts.

In earlier chapters, we have already seen the planning aspects of the I_{u-ps} interface and SGSN. However, there are some extra interfaces (i.e. G_n and G_a) that need to be planned apart from planning IP addressing and routing. As other interfaces – such as G_i, between the GSN and external networks – in the core network involve external element characteristics, they are considered to be outside the scope of this book.

GSN Inter-connection

GSN inter-connections (G_n interface planning solutions) are a part of the PC core network detailed plans. Traffic calculations are required so that traffic can be distributed equally between GGSNs. One way to send traffic is in a round-robin fashion whereby the traffic is sent to the next available GGSN. Another way could be a DNS deciding to which GGSN the traffic is routed through to the SGSN. Also, the amount of traffic is another factor to consider when deciding the routing for the traffic between the GGNs as it can be on an equal- or unequal-sharing basis.

G_a interface planning involves one more network element, the charging gateway. Information exchange or message transfer between the GSN and charging gateways is done using GTP-based protocols. Inter-connections between these network elements are planned based on the capabilities of the network elements and network configuration. One of the possible ways to inter-connect these elements is to use virtual LANs along with the planning of IP addressing, redundancy of paths and associated signalling etc.

9.4   CORE NETWORK OPTIMISATION

The optimisation process is quite similar to that seen in GSM/GPRS/EDGE core networks. Again, quality of services is an important backdrop against which the optimisation needs to be done (see Figure 9.10). Definition of key performance indicators (KPIs), analysis and preparation of the final optimisation plan are almost the same as what has already been discussed in Chapter 4 (for the CS core) and Chapter 5 (for the PS core). The final optimisation plan will consist of suggestions for changes to the existing network elements (e.g. increased capacity), new network elements and their locations, definition of allocation of transit elements and layers, new simplified number plans, etc. Also covered is optimisation

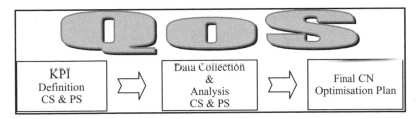

Figure 9.10 3G core network optimisation process

of the interfaces within the network, such as I_{u-cs}, I_{u-ps} and G_n, and interfaces to external networks, such as G_i.

However, one aspect does change with 3G networks, and that is the timing of the optimisation. With GSM networks, core network optimisation did not usually begin instantly after the network launch. It took some time (a few months to a few years) before it was realised that core network optimisation was needed, whereas the radio network optimisation cycle started almost in parallel with network planning. In some cases of 2.5G networks, and almost all cases of 3G networks, core network optimisation will be an integral part of network optimisation. The reason for this is that quality of service in 3G is considered from an end-to-end perspective, rather than from an individual network subsystem perspective (see Figure 9.11). As quality of the packet data is so important, the following section focuses on QoS of packet data and ways to improve it.

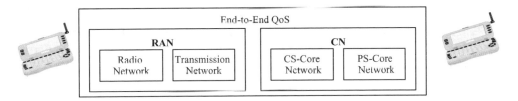

Figure 9.11 E2E QoS structure

9.5 END-TO-END QUALITY OF SERVICE

Mobile subscribers' perspective of the quality of the network is what matters, especially in third-generation networks where QoS is the backbone of design and implementation. All the three network subsystems (radio/transmission/core) contribute to the quality of services delivered.

For a given application, a group of parameters are identified and are called QoS parameters. One important aspect of optimisation is to keep this parameter group as precise and small as possible, because having too many parameters leads to increased complexity in the network and in the optimisation process. Each parameter group has control mechanisms, because the quality of any given application should be negotiable.

As we have seen, four traffic classes are defined in UMTS networks: conversational, streaming, interactive and background. Various applications fall under each category, but technically speaking these four classes can be differentiated by two factors:

- maximum permissible delay
- maximum allowed errors (BER).

These traffic classes and the QoS associated with them are responsible for handling packet flow (PDP contexts) in the network. The important aspects of PDP contexts include the type of PDP context, associated network elements (e.g. GSN in the case of a PS core network), and the expected QoS of a PDP context. Applications belonging to different traffic classes – WWW, Telnet, video conferencing, FTP, etc. – need different processes to achieve the desired QoS. Whenever a mobile subscriber activates a PDP context, the parameter group can be negotiated both in terms of quality (of service desired) and quantity (number of PDP contexts desired by the user). The network should be optimised so as to meet the most stringent demands made by the subscriber in terms of quality and quantity. This means that the correct parameters have to be assigned by the network to a given application and to the availability of the necessary resources. This loops back to the initial discussion of optimising the quantity of network elements and the interfaces between them.

In the initial phase of optimisation, when the traffic quantities may not be known, measurements can be made to observe the quality of the network. For example, for Web-related services, parameters such as the call (session) access rate and call (session) establishment time, dropped call (session) rate, etc., can be observed, which in turn are dependent upon PDP context establishment. Also, data capacities in the uplink and downlink directions can give an estimate of the end-to-end QoS of the network from the WWW perspective. Similarly, the quality of all applications important to the network should be measured, two of which are FTP and WAP file transfers.

End-to-end QoS is an area that is still being explored in third-generation networks. It will remain an area of focus for some years because each new application will need a different QoS.

IV

Fourth-generation Network Planning (OFDM/ALL–IP/WLAN)

10

4G Network Planning

10.1 INTRODUCTION TO 4G MOBILE NETWORKS

With the third-generation network deployment yet to pick up speed, fourth-generation technology is already in view. If the predictions of the mobile industry experts prove to be true, fourth-generation network deployment may start at any time in the coming decade. Trials have already being conducted by some mobile operators and vendors. But why is this 4G technology needed when 3G networks seem to be sufficient to cater for subscriber demands for high data rates and quality of service?

The answer is that present 3G capability is considered to be substantially less than predicted future requirements and applications. Also, future systems should be much cheaper for consumers. Thus, the concepts can be summarized as:

- Fourth-generation networks will provide subscribers with a higher bandwidth and a mobile data rate of 100 Mbps and more.

- It is expected that third-generation networks will not be able to meet the needs of services like video-conferencing, full motion video etc. in terms of QoS.

- There will be greater mobility and lower costs.

- It will be possible to integrate WLAN and WAN.

Moreover, fourth-generation networks will not be by-product only of the mobile industry!

The first research began around the early 1990s so as to develop technology that could cater for very high data rates, with simultaneous guaranteed QoS. The technology may see some peculiar features, such as cell phones operating in very high speed vehicles (e.g. trains running at more than 200 km/h). Present subscriber requirements include downloading videos and music etc., but the future seems to be moving towards applications like on-line games that demand immense capacity, greater QoS and very low costs! In short, a

Fundamentals of Cellular Network Planning & Optimisation A.R. Mishra.
© 2004 John Wiley & Sons, Ltd. ISBN: 0-470-86267-X

Table 10.1   Comparison of 3G and 4G network technologies

Key features	3G networks	4G networks
Data rate	384 kbps to 2 Mbps	20–100 Mbps
Frequency band	1.8–2.4 GHz	2–8 GHz
Bandwidth	5 MHz	About 100 MHz
Switching technique	Circuit- and packet-switched	Completely digital with packet voice
Radio access technology	WCDMA, CDMA-2000 etc.	OFDMA, MC-CDMA etc.
IP	IPv4.0, IPv5.0, IPv6.0	IPv6.0

4G system must be capable of providing highly efficient and cost-effective solutions for wireless network users.

Table 10.1 gives a comparison of few key features of 3G and 4G technologies.

10.2   KEY TECHNOLOGIES FOR FOURTH-GENERATION NETWORKS

Although there are a few technologies vying for the top stop for fourth-generation networks, OFDM and MC-CDMA may turn out to be the key competitors for the physical interface, and All-IP and WLAN for the upper layers. This section introduces OFDM (orthogonal frequency-division multiplexing) for the air-interface and All-IP for the upper layer. Later in the chapter, an overview of WLAN systems and network planning is given.

10.2.1   Orthogonal Frequency-division Multiplexing

OFDM is a frequency-division multiplexing technique that is used to transmit large amounts of data on a radio signal. Basically, a 'big' radio signal is subdivided into smaller signals and then transmitted to the receiver using different frequencies.

It is thought that OFDM will be able to fulfil the three most important requirements of 4G mobile networks: higher *coverage* and *capacity*, with desired QoS at minimum *cost*.

The biggest advantage of the OFDM technique is the mutual orthogonality of its carriers, which provides a high spectral efficiency. This is possible because there is no guard band and carriers can be packed very close together. Most of the alternative techniques require guard bands. In OFDM, even without a guard band, there is no interference because the carriers are orthogonal. The spectrum for OFDM lies between 200 MHz and about 3.5 GHz, with a spectral efficiency of about 1 bit/s/Hz.

Coverage in CDMA systems is limited by the phenomenon of *cell breathing* (described elsewhere in this book), as an increasing number of users decreases the area covered owing to an increase in interference. In an OFDM system, the *cell overlay technique* is used (similar to that in GSM), thereby reducing co-channel interference.

Network planning for an OFMD system is quite similar to that for GSM/GPRS. This is because frequency re-use is reintroduced (unlike in WCDMA, where the frequency re-use factor was 1, theoretically). For this reason, the power control feature in OFDM networks is

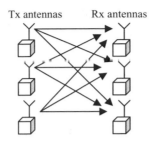

Figure 10.1 Capacity increase using MIMO antenna systems

not as essential as in WDCMA networks. In WCDMA radio networks, power control and spread spectrum are required for reducing interference. In OFDM radio networks, accurate estimation of frequency offset is required.

Increasing the number of transmitting and receiving antennas can increase capacity. Multiple-input/multiple-output (MIMO) antenna systems can be used, as shown in Figure 10.1.

Network planning for OFDM networks is simpler than for CDMA networks. OFDM reduces the amount of crosstalk in signal transmissions. Thus, in a nutshell, we can see that OFDM clearly has an edge over CDMA, making it the preferred air-interface technology for future mobile networks.

10.2.2 All-IP Networks

Structure

The All-IP network has been tipped as the most probable technology to be synonymous with fourth-generation networks. A simplified All-IP network is shown in Figure 10.2.

The most important difference between the All-IP network and existing 2G and 3G networks is in the functionality of the RNC and BSC, which is now distributed to the BTS and a set of servers and gateways. Various elements in this network are described below.

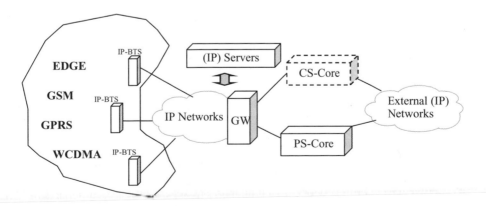

Figure 10.2 Example of an All-IP network

- IP-BTS: The functionality of the IP base station in this network is more than the functionality of base stations seen in earlier chapters. This base station performs also as a mini-RNC/BSC, generally capable of performing layer 1, 2 and 3 functions. There are two types: serving BTS and drift BTS (equivalent to serving RNC and drift RNC in a WCDMA radio network).

- (IP) servers: The IP base station is not capable of performing all the RNC/BSC functions, which are of network level. These servers handle the signalling between the network elements. They are capable also of auto-tuning the parameters of the radio network, leading to better utilization of radio resources. As there are multiple technologies to be handled, a common server improves the performance and efficiency of the network in comparison with separate servers for each of the radio interfaces.

- Gateways (GW): These are responsible for the interaction of the IP-RAN and IP-Core networks. They are usually of two types, CS-GW and PS-GW, based on the type of call (circuit-switched or packet-switched) it is capable of handling.

Network Planning for the All-IP Network

Network planning covers the access (transmission) network and the packet core network. Figure 10.3 shows a small box with the core network as a subset of the packet core network, indicating that voice traffic will still be a part of mobile communications, but it will travel on the packet core network (as opposed to the circuit core in 2G and 3G networks).

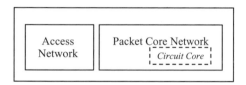

Figure 10.3 Network planning for an All-IP network

Process and Protocol Overview

The transmission network planning process is similar to that discussed in Chapter 8. The process thus starts with dimensioning and pre-planning, followed by detailed planning and implementation. The main steps in pre-planning will be:

- dimensioning the number of network elements such as IP base stations, servers, PS- (and CS-) core network elements etc.

- dimensioning the capacities of 'open' interfaces

- tackling inter-operability issues between the GSM/UMTS/WLAN networks.

Transmission and core network planning are more dependent on each other compared with 3G network planning. As the data traffic will be higher in quantity and quality, delay study will constitute an important part of the planning process.

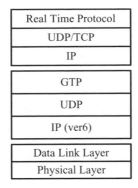

Real Time Protocol
UDP/TCP
IP

GTP
UDP
IP (ver6)

Data Link Layer
Physical Layer

Figure 10.4 Protocol structure in an All-IP network

The major change in the transmission network is the use of All-IP for the flow of traffic. Third-generation transmission networks use an ATM (asynchronous transfer mode) layer for the flow of traffic, while an All-IP network will not have an ATM layer (see Figure 10.4). The major impact of this on transmission network dimensioning is reduced overheads. Overheads in the data link layer will depend on the media. The IP (version 6) layer takes over from ATM layer in these networks.

10.2.3   Wireless Local-area Networks

Performance

A wireless local-area network (WLAN) is a flexible data communication system, being an alternate to existing wired LANs. This technology removes the hassel of taking wires and cables to and from equipment in an office environment. The Institute of Electrical & Electronic Engineers developed the standards for WLAN, specified in IEEE 802.11. The initial standards specified an operating frequency band of 2.4 GHz and a theoretical data rate of up to 11 Mbps. Subsequent issues of the standards have increased the capacity of the WLAN to 54 Mbps, in the same frequency band.

What is the expected role of WLANs in fourth-generation networks? It is expected that WLANs will complement the existing 3G and All-IP 4G networks in high-density area networks by providing similar services at an even higher bandwidth (compared with mature 3G and 4G radio networks). The technology is considered to be best suited to low-usage mobile users who want high data rates in an indoor setting. This also means that the mobile equipment should have the flexibility to choose the access technology at any given time, depending on the environment.

Network Planning for a WLAN

Network planning is expected to focus mainly on indoor coverage. The principal aspects of the pre-planning phase are:

- the area for which the network is planned
- subscriber database information

- coverage and capacity requirements

- the number of channels that can be used (e.g. 13 in Europe and 11 in the USA)

- the number of channels that can be used simultaneously without interference

- propagation conditions (e.g. multipath in an indoor environment)

- equipment data (e.g. antenna gains and transmitted power)

- the air-interface radio link budget.

Based on these factors, coverage, capacity and quality can be calculated.

A WLAN network should provide coverage for 100% of the area for which it is being planned. Most of the issues that we have seen in earlier chapters on radio network planning – such as coverage threshold, signal quality, C/I etc. – will be involved in planning coverage, but with more stringent requirements. Methods to improve coverage include increasing the power levels, incorporating diversity schemes etc.

Frequency planning is another area of challenge in WLAN networks. As the number of channels available is less than the capacity likely to be demanded, frequency planning becomes a crucial task. As frequency re-use becomes lower, the quality (and hence through-put) becomes lower. One rule of thumb could be to plan these networks with a frequency re-use of more than unity.

10.3 CHALLENGES IN 4G WIRELESS NETWORKS

Two main challenges need to be addressed before fourth-generation networks become a reality. The first concerns accessibility to different types of cellular network. The second concerns how to maintain the desired end-to-end QoS for traffic that has varying require-ments of bandwidth, bit rates, channel characteristics etc., and especially the handover delays, which are a cause of worry. During the handover process, mobile subscribers are expected to face a drop in the QoS level.

Fourth-generation networks are still 'unclear' from the perspective of defining the net-work planning processes. This is partly because the evolution of these 4G networks is not driven only by the mobiles industry. Moreover, standards-defining bodies like the IEEE are still in the process of producing standard recommendations. A major challenge is the planning of handovers not only between different generations of networks but also between different technologies of the same generation (All-IP to WLAN etc.), *while maintaining the QoS standards*.

Appendices

Appendix A
Integrated Network Planning Tool: Nokia NetAct Planner

Ari Niininen
Nokia Networks

A.1 OVERVIEW OF NETACT PLANNER

NetAct Planner is a multi-user environment planning tool based on a Windows® platform. It has several interfaces to other planning systems, measurement tools and the Nokia NetAct network management system for deploying plans into the network and back into the planning tool, as shown in Figure A.1. Table A.1 provides a summary and is displayed at the end of this section.

A.2 THE RADIO PLANNER

Radio Planner is an efficient and flexible solution for radio network planning. It offers effective support in all phases of the radio network planning process for multi-user organisations in an office environment. Furthermore, PC compatibility with laptop computers enables field use and provides planning engineers with real mobility.

Radio Planner is specially developed for the planning of all types of GSM and TETRA networks. Radio Planner was designed using a systems approach, which helps to provide

Fundamentals of Cellular Network Planning & Optimisation A.R. Mishra.
© 2004 John Wiley & Sons, Ltd. ISBN: 0-470-86267-X

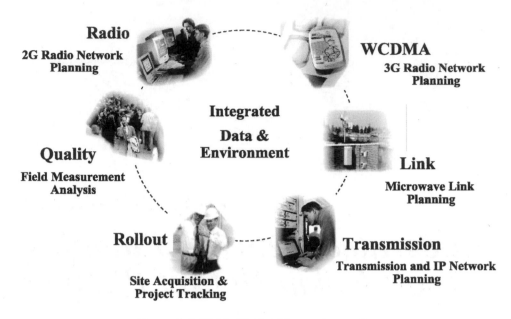

Figure A.1 Nokia NetAct Planner integration

mobile subscribers with continuous coverage, capacity, quality and service content. Radio Planner supports standard GSM features such as frequency hopping, dual band as well as Nokia GSM system specific features.

GPRS and EDGE planning are key challenges for many operators. Radio Planner supports packet-switched radio network capacity planning. With Radio Planner, you can calculate the needed GPRS and EDGE capacities in the network based on the selected quality of service criteria and data rates.

Coverage can be calculated with the four included propagation models: Okumura–Hata, a Nokia version of Okumura–Hata, Walfisch–Ikegami, and Microcell. There is also an open interface to external propagation models, which means that you can utilise your own propagation models if needed. All of these models can be tuned with drive test measurements. Radio Planner also provides coverage planning and analysis support for dual-band networks (shown in Figure A.2).

Capacity calculation is based on real network traffic imported from Nokia NetAct (Network Management System). You can distribute this real network traffic data on user-definable clutter or vector weights, enabling accurate capacity calculation, which is especially important in dual-band and GPRS/EDGE planning.

The frequency allocation tool minimises interference efficiently by optimising the separation cost and interference table cost. Allocation accuracy is based on allocation time. The progress of allocation can be followed with statistical or graphical views. The frequency allocation tool supports both RF and base-band frequency hopping.

Information management is becoming increasingly important in radio network planning, especially in projects such as microcell planning, where the major part of planning activities can be carried out in the field. Radio Planner has a user-friendly site database, which enables

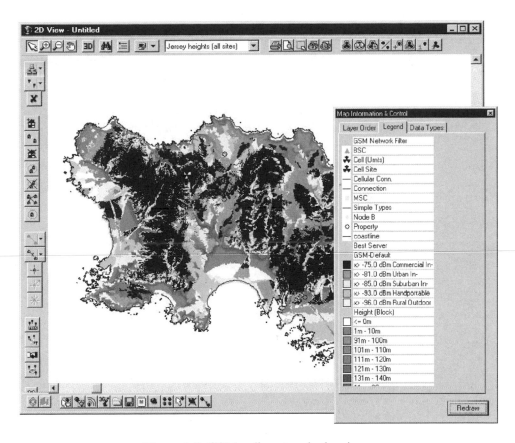

Figure A.2 GSM radio network planning

fast entering and editing of all network parameters. Radio Planner reporting tools provide flexible and extensive network reports.

A.3   WCDMA PLANNER

WCDMA Planner offers excellent assistance with radio network planning when a 3G Wideband Code Division Multiple Access (WCDMA) network is built. It is a flexible and user-friendly planning system capable of meeting your initial WCDMA radio network planning needs.

The WCDMA planning process in WCDMA Planner is designed to be very fluent and easy to follow (see Figure A.3). Furthermore, powerful editing and data handling features make the whole network planning process faster and easier. All this results in superior usability.

In addition to high-quality radio network planning features such as advanced and tunable propagation models, WCDMA Planner has a revolutionary approach to planning and optimising WCDMA networks based on UMTS service and traffic requirements. WCDMA

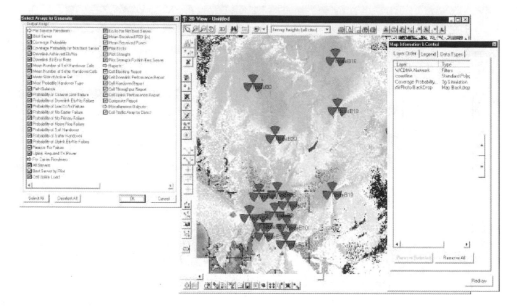

Figure A.3 WCDMA radio network planning

Planner supports the modelling of several services and their mixtures through flexible traffic and service modelling tools.

The WCDMA Planner traffic modelling function provides you with various methods for defining traffic forecasts and for the effective management of traffic data via traffic layers. The realistic modelling of non-homogenous traffic distribution information substantially improves the accuracy of the detailed analyses. WCDMA Planner uses state-of-the-art simulation techniques such as Monte Carlo to estimate the interference in the system.

WCDMA Planner comprises several analyses to provide an in-depth understanding of the plan. For example, you can calculate the probability with which each service can be provided, or visualise the area in which a soft handover occurs. Analysis results are displayed in tables, reports and as map displays that can be printed.

NetAct Planner enables common planning for 2G and 3G networks. As WCDMA Planner is part of the Nokia WCDMA system solution, it is the most suitable tool for planning Nokia radio access networks, along with networks from other vendors.

A.4 LINK PLANNER

Link Planner is an efficient and flexible solution for microwave link planning. Link Planner provides line-of-sight (LOS) checking, link frequency allocation, interference analysis, advanced link database and a traffic routing tool. Link Planner calculates link budgets, outage, availability and reliability values (see Figure A.4).

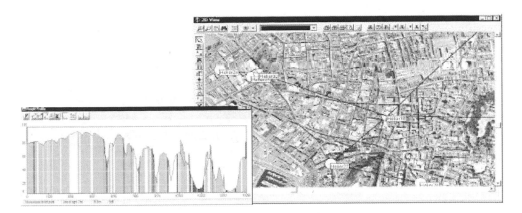

Figure A.4 Microwave link planning

Height profiling is able to perform LOS checks between: site to site, site to point, point to point, point to multi-point. It also does calculations related to Fresnel zones, reflections, minimum antenna height, clearance/obstruction, hop length, diffraction loss, etc. Link Planner supports ITU-R recommendations (P.530-7 and P.530-10) and ITU-T recommendations (G.821 and G.826) performance estimations. Also it supports diversity cases: space, frequency, space + frequency and angle diversity.

Interference analyses are done using a module called an interference wizard. It performs calculation such as cumulative analyses of interference, co-channel and adjacent-channel interference. Also part of the module are preparation of related reports, visualisation of the interference in a 2D view, its subsequent effect on link budget calculation, etc.

Link Planner shares a common database with other modules of NetAct Planner, which enables seamless data usage between Link Planner and other NetAct Planner modules. This means that both the transmission and radio network planning teams have real-time information of each other's project work, and both teams can synchronise different planning project phases smoothly.

A.5 TRANSMISSION PLANNER

Transmission Planner is a complete transmission network planning tool solution with sophisticated, easy-to-use information management functionality, routing facilities, and manipulation of physical and logical networks as well as capacity and equipment type and size identification accordingly.

Transmission Planner provides advanced support for planning and dimensioning of mobile, TETRA, PSTN and Data Operator's networks. SDH, primary rate ($n \times 2$ Mbps) and generic data transmission as well switching and SS7 signalling networks planning are supported.

Mobile access transmission network planning ($n \times 2$ Mbps) is supported in all details. Transmission Planner combines sophisticated design algorithms with an intuitive, easy-to-use graphical user interface. Design algorithms are useful but they cannot handle all

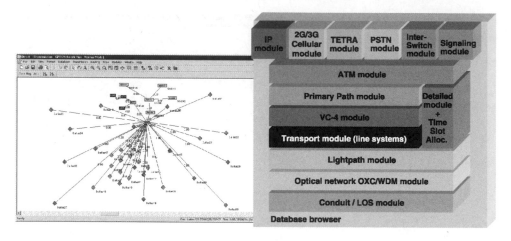

Figure A.5 Capacity planning and modules in Transmission Planner

complexities of real networks. A user-friendly interface is important for visualisation and for manual modifications to provide maximal flexibility for the network planner.

The Transmission Module supports efficient transmission network architecture planning in interactive sessions. Applications are available for key tasks, such as network topology and capacity manipulation with different traffic scenarios. It gives results of network capacity utilisation as well as information of non-routed traffic. As a result, the network topology and required transmission capacities are presented, as well as the number and size of nodes (see Figure A.5).

The Transmission Module's Detailed-view provides a detailed level of information about the actual physical devices in 2G and 3G transmission networks. Other modules in transmission network layers are: IP, ATM, Lightpath, SDH, and PDH. The Traffic Module models all traffic for mobile networks.

The Traffic Module is a logical layer with visible nodes including: BTSs, BSCs, MSCs and their 3G equivalents and their logical connections. Other service network layers are: TETRA, PSTN Access, Interswitch, and SS7 signalling.

The main driver for creating IP and ATM modules is the increasing importance of packet-based networks. Most of the modern data services are implemented in the various packet-switched networks and the IP module was created to provide a means of modelling these. In addition to well-established packet networks that are based on IP or ATM packet networks, Nokia's aim is to provide a basic transmission capacity dimensioning tool both for GPRS networks and eventually for third-generation mobile networks.

The IP module is used to define traffic demands for an IP network. As planning results, equipment capacities and estimates of the network usage are provided. The IP module can provide valuable information about bottlenecks in the network and easy ways to solve them with the user-friendly graphical interface. It provides an easy way to estimate packet throughputs; and to give more realistic results, packet loads of equipment and protocol overheads can be included in the capacity calculations.

IP traffic can be easily routed to lower transport layers such as ATM, SDH or PDH and the required capacities on those layers can be analysed in corresponding modules.

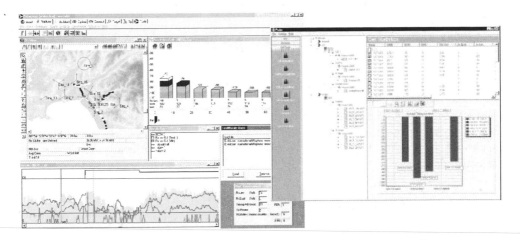

Figure A.6 Drive-test results analysis using Quality Planner

A.6 QUALITY PLANNER

It is essential that mobile network operators measure the performance of the radio network and carry out optimisation of the system. Only by making field trial measurements are operators able to determine the performance of the radio network as seen by subscribers. Quality Planner provides an automated analysis of the test mobile measurement files and produces diagnostic reports to resolve problems within the radio network (see Figure A.6).

Quality Planner is an efficient and flexible solution for analysing GSM network air interface measurements, providing effective support in all analysis tasks including GPRS and EDGE network measurement analysis, for network planning organisations. PC compatibility with laptop computers enables field use and provides planning engineers with real mobility.

Quality Planner has a common database with other NetAct Planner modules, thereby enabling seamless data usage between radio planning and optimisation. Drive test measurements can be analysed in the same map view as coverage, traffic or interference predictions, enabling a unique and accurate network analysis. Quality Planner also offers efficient problem solving and diagnostic capabilities.

The drive test replay functionality of Quality Planner enables a broad range of information to be displayed in the 2D map window during the drive test measurement analysis, including layer 3 messages, serving cell information, radio environment data, serving and neighbour graphs.

Quality Planner supports a thorough analysis of radio network performance. The call success analysis is an important part of the field test measurements. This function summarises the status of each call made within the field trial measurements. Handover analysis provides the ability to analyse cost-effective methods for both inter- and intra cell handovers. The neighbour analysis performs a checking routine based on the real BSC neighbour lists. Possible missing or not measured neighbours are registered as problems. The signal level analysis clarifies the network field strength levels and the quality analysis co- and adjacent-channel interference levels.

Quality Planner's Probe module is a network analysis tool that enables you to benchmark the quality of your network and easily compare it against your own network's past performance.

Whereas the analysis functionality of Quality Planner is designed to pinpoint specific problems and faults in the network so that you can locate and fix them, Probe gives a wider picture of one or more networks, providing statistics that show the system's problems as a whole. By using Probe, you can also perform trend analyses on key performance indicators such as handover, level, quality and network performance. Additionally, you can query selective data.

A.7 ROLLOUT PLANNER

Rollout Planner is a project management system for telecom network rollouts and expansions. It handles the project management for implementation projects.

Rollout Planner is a complete, efficient and flexible solution for project management and site tracking. The Rollout Planner platform is suitable for multi-user projects and organisations; it also supports field use and therefore provides real mobility for users (see Figure A.7).

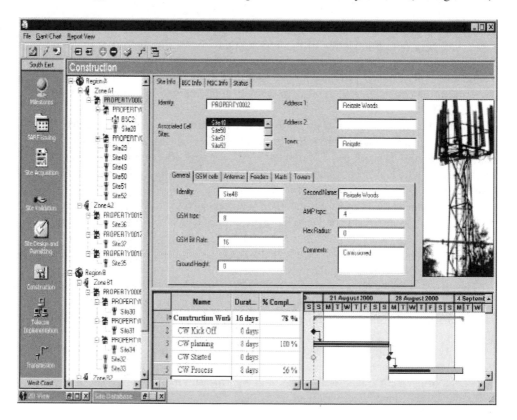

Figure A.7 Rollout Planner

Rollout Planner provides advanced support for project management, including:

- detailed information at site level
- planning of project milestones and tracking of different tasks
- management of site acquisition and site candidates
- status information of the project on a per-site basis
- management of task-related documents
- productivity and deviation reports on individual or group of sites
- control of lead times between tasks and milestones.

Rollout Planner has a common database with other modules of NetAct Planner, to enable seamless data usage between project management and network planning, including site acquisition and implementation status. This means that the project management and network planning teams have real-time knowledge of each other's project work and all departments can smoothly synchronise different planning project phases.

Table A.1   Summary of NetAct Planner

Module	Functionality
Radio Planner	Radio Planner is developed especially for the planning of all types of GSM and TETRA networks and is also suitable for NMT networks. It supports present GSM features such as frequency hopping and dual-band planning. HSCSD, GPRS EDGE technologies are also supported.
WCDMA Planner	WCDMA Planner is designed for WCDMA radio network planning. This planning system is part of the Nokia WCDMA system solution and has been developed in cooperation with both Nokia and a leading planning tool supplier, therefore enabling the latest WCDMA technology utilisation. In addition to high-quality radio network planning functionality, WCDMA Planner offers an evolutionary approach to planning and optimising networks based on data service requirements.
Link Planner	Link Planner is an efficient and flexible solution for microwave link planning. Link Planner uses the most modern ITU formulas and includes interference calculations for frequency assignment.
Transmission Planner	Transmission Planner is designed for transport and switching network planning. It supports the planning of 2G/3G mobile, ATM/IP and PSTN networks, including dimensioning and network architecture comparisons.

(continued)

Table A.1 (*Cont.*)

Module	Functionality
Quality Planner	Quality Planner for field measurement analysis is an intelligent software package that provides automated analysis of test mobile measurement files and produces diagnostic reports to resolve problems within the radio network.
Rollout Planner	Rollout Planner is a site acquisition and project process-tracking tool for network rollout and expansion projects, providing full visibility of project milestones, progress and productivity.

Appendix B
MMS Network Planning

Christophe Landemaine, Carlos Crespo

Nokia Networks

B.1 INTRODUCTION TO MMS

MMS (multimedia messaging) is an end-to-end application for mobile messaging, mobile-to-mobile, mobile-to-Internet and Internet-to-mobile. It will provide rich multimedia content, including images, audio, video and text. It is a globally standardised service and happens to be one of the first 3GPP standardised 3G services. In principle, MMS content can include one or several of the following content types, with minimal restrictions to message size or format:

- picture

- text

- audio

- video.

To create a multimedia message, a terminal with a connected camera is used for the initial digital image input. If a multimedia message is sent to a mobile phone without the MMS application, this so-called legacy terminal will receive an SMS stating that at the operator website there is a new message along with the Internet address and personal password to fetch it.

Fundamentals of Cellular Network Planning & Optimisation A.R. Mishra.
© 2004 John Wiley & Sons, Ltd. ISBN: 0-470-86267-X

Figure B.1 SMS versus MMS

The basic difference with respect to the well-known SMS is that the message can be a combination of text, pictures and voice, in a roughly similar way to e-mail and attachments. This gives the first major difference from SMS, where the size of one message is limited to 160 characters. Now the size of the MMS is variable, ranging from a few kilobytes to tens of kilobytes in most of the cases (see Figure B.1).

This service and its associated infrastructure should be integrated into the existing network. This document will concentrate on the MMS delivery through GPRS, which will be the main expected bearer during the early stage, although it is also possible to utilise the CSD as a bearer, and 3G as well.

B.2 MMS SOLUTION AND ARCHITECTURE

The generic MMS architecture as defined in the standards is presented in Figure B.2.

It is important to notice that 3GPP standards define an MMS server for storing and handling and a MMS relay is responsible of the transfer of messages between different

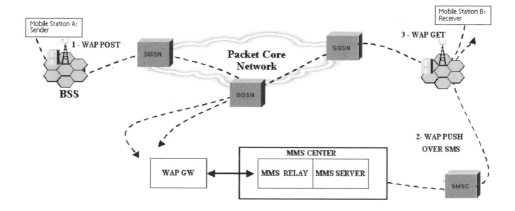

Figure B.2 General overview of MMS system architecture

messaging systems. On the other hand, the WAP-Forum organisation defines an MMS server and MMS proxy-relay. In most cases these two functions (MMS server and MMS relay) are combined and we commonly call the unit an MMS server or MMS centre.

B.3   MMS NETWORK PLANNING

MMS Network Dimensioning

The input requirements for proper dimensioning of an MMS-capable GPRS network are described below. Figure B.3 shows a general flow for MMS dimensioning and planning.

Traffic Input

Traffic-related input is probably the most important from a planning point of view. It can be categorised according to different criteria. The following sections provide a deeper understanding of the different types of traffic input that can be given.

Current GPRS Network Assessment
It is crucial to understand the existing GPRS infrastructure, since MMS will be deployed over existing GPRS networks in most cases. This assessment must be done from several points of view with the existing traffic load and configuration, and in particular:

- SGSNs: number, HW and SW version, capacity equipped, interfaces types and capacity

- GGSNs: number, HW and SW version, capacity equipped, interfaces types and capacity

- DNS: firewall, WAP gateway and router configuration and capacities

- core sites: geographical location, backbone connectivity, centralisation/distribution of servers and other equipment

- types of service offered: Internet access, corporate intranet access, e-mail, WAP

- current level of usage of those packet services, including traffic load for the GSNs and the backbone.

The output of this assessment can be a report containing a description of the above items as well as a graphic diagram of the current packet core network, preferably showing the VLAN level as well. Availability of this report (or equivalent information in several formats) is the starting point of the MMS planning activity.

Traffic Mix (Size and Content)
The concept of traffic mix in this context is more related to usage pattern and MMS service penetration. Another difference to consider with respect to SMS is that now we may have more than one MMSC involved in delivery, for example in the case of a distributed core network with MMSC in two or more sites, regardless of the clustering strategies.

The basic traffic input for MMS can be further described as:

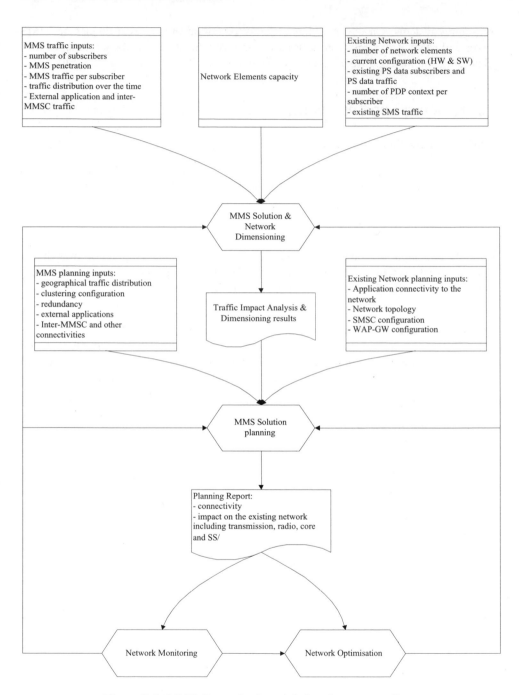

Figure B.3 MMS dimensioning and planning process flow

- total number of subscribers and MMS penetration

- estimated number of MMS per subscriber per busy hour (different usage data can be given, for instance, MMS per user per month)

- GPRS penetration, in own network and other PLMNs if available

- average size of MMS in kilobytes (a good reference value to start with could be 30 kilobytes, but any other combination or distribution can be taken into account).

The following GPRS input values should also be considered:

- number of PDP context per subscriber

- total existing PS traffic per subscriber per busy hour

- number and topology of BSCs

- additional BSS information, like number of RA, number of LA, etc.

Traffic inputs can be given in a table, as in the example in Figure B.4.

Traffic Distribution over Time (BH/Burstiness Considerations)
The impact of the burstiness and load on busy hours has also to be considered. There are few possibilities to define the traffic behaviour over time, and the network planner needs to be aware of them to properly understand the input and provide the adequate output. For product dimensioning purposes it has been defined as:

- number of MO and/or MT MMS per user per month (it may include different access to different servers included in the concrete MMS service concept)

- number of busy hours per day, when the traffic distribution is provided so that the busy period of the day for MMS traffic extends to more than one hour

- peak traffic multiplier, when this peak correction factor is included to enable the peaks of traffic.

Input Data					
	2002	*2003*	*2004*	*2005*	*2006*
Number of subscribers	1 500 000	1 600 000	1 700 000	1 800 000	2 000 000
Number of PDP context per user per BH	1	1	1	1	1
Per subscriber's PS traffic in BH (kByte/h)	100	120	150	180	200
MMS size (kByte)	30	30	50	70	100
Number of MMS per user per month	20	30	40	45	50
Peak hour multiplier	4	4	4	4	4
Number of MMS per user per BH	0.11	0.17	0.22	0.25	0.28
GPRS Penetration	13%	34%	55%	65%	75%
MMS penetration	25%	50%	75%	95%	95%

Figure B.4 MMS traffic inputs example

It is also important to consider, when possible and applicable, the current traffic distribution of SMS, with measurements taken from an existing network, if they are available. This is a valuable source to complement the forecast provided by the customer, although it should be considered case by case.

Geographical Distribution

Generally, a good understanding of the network topology is a requirement, particularly when defining locations of servers and distribution strategy. In order to better identify the sites for deployment, the traffic volume of the urban areas and the routing areas definition on the GPRS network should also be understood correctly.

Once more, previous experience of SMS traffic and its geographical distribution may be a help. An increase in the number of MMS coming from and going to a different PLMN can be expected, particularly with operators from different countries. The MMS will play a role similar to that of traditional postcards for tourists on holiday. This can be taken into account when studying the roaming strategy.

Subscriber Growth Estimates

This is the key input in terms of definition of a rollout plan for the MMSC, and its impact in the expansion of the existing GPRS. It can be expected to be provided from customer data, usually in the format of a table as shown in Figure B.4.

MMS Configuration Issues

The MMSC solution and servers can be defined in various formats. Therefore, MMS configuration plays a major role in the overall planning process. Site connectivity is also heavily affected by the integration of MMS.

Site Definition

The MMS solution is to be integrated within the core site, and connectivity has to be defined both at both physical and logical levels. A cluster configuration involves additional complexity for connectivity. The server or cluster has to be integrated into the Gi VLAN. The connectivity to the Gi is created through the associated WAP gateway.

Since the MMS system connectivity is to be established at the internal Gi interface, we can assume that there could be a requirement for physical connectivity to the local site switches.

If the system presents a cluster solution, the corresponding load balancing switch can be placed before connecting the switch.

Redundancy and Load Balancing

The redundancy of the solution depends heavily on the redundancy of the servers. A cluster architecture for larger systems enables more options for physical connectivity redundancy. The use of load balancing switches is another of the key issues to be considered. From a planning point of view, this can be considered as an input, defined by the configuration offered in every case.

External Application Gateways and Services

There is a need to establish and define connectivity between the MMSC and other servers which provide applications. Since every project will deploy different external applications, it must be defined case by case. VLAN separation offers the possibility of easier management and operation. The addressing plan should take into account the required allocation of IP addresses in both cases, when using VLAN or when using direct connections.

SMSC

SMSC should be considered part of the MMS solution, since it plays an active role in the MMS delivery process. SMS is the bearer for the transmission of notifications and delivery reports to subscribers. It is considered that the delivery of a 'mobile terminated' MMS within a network will generate additional SMS in the network. Thus, the plans and capacity forecasts of an SMS system must be reviewed in line with the penetration of MMS-capable phones.

WAP Gateway

The main connection is towards the WAP gateway, which will provide the WAP mechanism for MMS delivery. The MMS system relies on the existing WAP gateway to provide connectivity to the GPRS packet core and push functionality, for instance. An adequate MMS configuration should be set in the WAP Gateway.

Inter-MMSC Connectivity and Roaming

It can be considered that, where two PLMN are involved, there are also two MMSC involved in a mobile-to-mobile delivery process: the sender uses its own network MMSC and the receiver obtains the MMS from its own network MMSC. Sender and receiver may have different kinds of services provided in different ways, if they belong to different networks. If sender and receiver belong to the same network, the physical MMSC involved in both processes is likely to be the same, although in larger networks there could be more than one MMSC, thus required additional WAN networked connectivity.

In terms of connectivity, several networking options are to be discussed, depending on the customer's availability and/or needs. It can be, for instance, a dedicated leased line, a VPN connection, or any other secure tunnelling solution, GRX, and even Internet – though this is not recommended for obvious security reasons.

Dimensioning of the MMS Solution

This can be considered as a traditional case of network dimensioning, where all the network links are supposed to be dimensioned according to the needs for the expected MMS traffic. This section will mostly describe and comment on the process as it can be performed.

Dimensioning of MMS System Servers

It is important to note that the actual dimensioning of the MMSC and servers is in most cases a fairly simple exercise, with product dimensioning based on a software license scheme of a maximum number of given transactions per second. It has to be checked that the hardware platforms of the MMSC and additional servers are able to deliver the service as per license agreements. But since the traffic is anyhow going through the whole GPRS network, the key of the dimensioning exercise from the network point of view is in dimensioning both radio and core network elements, including their interfaces. These values are directly calculated from the input provided by the customers in RFQ or similar tendering documentation, and they are the result of product dimensioning required to handle the expected MMS traffic. That MMS traffic handled by the MMS system is the key of the overall network dimensioning exercise.

Rollout Network Planning according to Subscriber Growth Forecasts

The dimensioning process follows the traditional approach of a year-based evolution, since this is the easiest way to present the MMS dimensioning within the context of the general rollout plan of the network. The network planner needs to get input in some of the traditional fields, often provided by tendering documentation. However, sometimes, because of missing inputs, new assumptions are needed to perform the dimensioning calculation. The input data considered for MMS services and which should be taken into account together with the general dimensioning sheet for PS traffic are given below.

The most important input can be the *total number of subscribers per year* – usually given as the end-of-the-year number of subscribers. This input is combined with the relevant penetration values: *GPRS penetration* in the most likely case of MMS over GPRS; and *MMS penetration*, given as the proportion of GPRS terminals (GPRS penetration) that are MMS-capable. Others inputs are:

- *Number of MMS messages per user per month*: Other possibilities may have to be taken into account.

- *Number of PDP contexts per subscriber per busy hour*.

- Although not so relevant for the purpose of MMS dimensioning itself, the existing PS traffic could be included in the overall calculation.

- The average size of the MMS message: It is expected that it will grow in the future, when not only pictures and text, but also short audio and video clips come into use.

- *Peak-hour multiplier*: This is introduced to model busy traffic behaviour.

- G_b overhead and burstiness factors, G_n overhead factor: These are to be taken into account in the respective interface calculations.

With these values, the result can be obtained in terms of *number of MMS per subscriber per busy hour*, which is the basis of the dimensioning exercise. MMS traffic just becomes some additional packet data traffic and the network needs to be dimensioned accordingly considering this new type of traffic.

Mobile-originated, mobile-terminated, application-originated or application-terminated MMS should be distinguished. For the GPRS bearer, and excluding the combination AO-AT, every event, either send or receive is a transaction, or a traffic happening, and so has to be taken into account. For each mobile-originated or mobile-terminated message, one PDP context needs to be activated.

Latest network element capacities should be checked and taken into consideration since it is possible to face different scenarios with different releases of these network elements. For example, throughput and number of PDP contexts are usually the main limiting factors when dimensioning SGSN and GGSN.

In this context, from the network planning perspective, the MMS is a key contributor to GPRS traffic, and later on to UMTS PS traffic. Dimensioning of MMSC and related servers is only part of the planning exercise. Good understanding of the impact of MMS traffic on the underlying network providing the bearer for the service – be the bearer GPRS or UMTS – is key for a successful MMS deployment and operation. Thus, a new reason for optimisation and performance enhancement arises, specifically for MMS or generic optimisation including MMS.

Networking: LAN Switching Capacity, IP addressing and Connectivity

Once the MMS traffic is identified and the traffic model is understood, the next logical step is to integrate the MMSC into the GPRS packet core network. It will be assumed for the sake of simplicity that the rest of the servers around the MMSC will not have direct interfaces with the GPRS packet core network. Inter-MMSC connections are dealt with later on. Regardless of that, the MMS traffic should be also taken into account for the capacity of the site switches.

Since the G_i connectivity is through a WAP gateway, IP addressing can be solved without interfering with packet core addressing. Should the site be equipped with a Cisco OSR or similar, a separated VLAN (MMS VLAN) seems to be a reasonable option. Addressing will depend greatly on the configuration of the server and/or cluster.

Redundancy and Load Balancing Recommendations for Servers

When there is more than one MMSC server, load-balancing solutions should be used to ensure even traffic distribution.

Inter-MMSC Traffic Analysis

Unlike with SMS, in MMS there is the possibility of having inter-MMSC traffic. In the case of SMS, once the message is received in the SMSC, it is delivered through the SS7 network to the destination. With MMS, if the origin and destination of the MMS are under different MMSCs there is a need to calculate the traffic. In the case of different PLMNs, inter-MMSC traffic can be assumed as one of the components of the overall GPRS roaming traffic, depending of course on the roaming agreements in place. Inter-MMSC traffic can be provided through an SMTP protocol. The interconnection can be carried out also through the Internet, although VPN or at least some type of tunnelling protocol must be used, as mentioned in section B.2. The dimensioning of the inter-MMSC traffic can be worked out as a traditional Excel-based exercise through the estimation of usage patterns, in a similar fashion to the SS7 connectivity requirements for inter-PLMN SMS dimensioning. In the early stages of deployment, with low traffic volumes, it should not be a bottleneck for the end-user experience.

B.4   EFFECT OF MMS ON GPRS/GSM NETWORKS

This section gives a general overview of the effects of MMS deployment in the existing network infrastructure. The effects can be described at two different levels: connectivity and traffic. Traffic impact is the major concern, since MMS data should not adversely affect the existing voice, messaging and packet data services. This section provides a first approach to the problem, and hints for analysis.

Radio Network

From the radio access point of view, MMS utilises the GPRS bearer. The dimensioning of the GPRS traffic for the air interface should include the MMS forecasts, or it should be modified to include them. It is not expected to have a great impact in the early stages, since traffic volumes should not be excessive while the penetration is low, and since the nature of the MMS traffic allows a certain delay without affecting the end-user experience. Thus, it can be considered as additional packet data traffic.

Cellular Transmission Network

From the cellular transmission point of view, the MMS traffic does not add concerns, other than a revision of the dimensioning and checking of the spare capacity of the links against the traffic growth forecasts of GPRS traffic.

SS7 and SMSC

Since the notification of the MMS will be provided through SMS, the SMSC is an integral part of the MMS system. It needs to be connected to the MMSC. Furthermore, the extra amount of SMS generated by the MMSC needs to be analysed and calculated. From the connectivity point of view, it is important to note that the SMSC needs to interface the MMSC or cluster. There is a need to take into account the extra traffic on the SS7 network due to notifications and delivery reports of MMS. Overloading of the SS7 network may significantly affect voice services, so a careful study of the SMS load should always be done. Frequent updates, when the real values can be measured, enable more accurate forecasting of the needs and can be of great help in deciding on the need of additional capacity.

WAP Gateway

As mentioned earlier, the WAP gateway provides the push functionality for MMS, and is the link between the MMS system and the G_i interface. The packet core connectivity is provided through the WAP gateway.

Core Network

As mentioned earlier, MMS traffic is seen as some additional packet data traffic from the core point of view. The main consequence of MMS is to bring more data traffic in the network and the need for PDP contexts for each MMS. Throughput and PDP contexts being the two main limiting factors of core dimensioning, the impact of MMS on the core will not be negligible and the main elements to be affected will be SGSN and GGSN.

One way to measure the impact of MMS traffic is to combine the results of PS data and MMS traffic dimensioning. When making the dimensioning for PS data, without MMS, we end up with a certain number of needed SGSNs and GGSNs and their configurations. A similar dimensioning exercise, now including MMS traffic, will show the real impact of MMS in term of needed core network elements. It is very important to analyse MMS traffic, particularly from the point of view of the growth rate of generic PS traffic versus MMS traffic, and to determine the key breaking points when more radio capacity is required or when SGSN/GGSN need to be integrated. Therefore, there should be a close follow-up of the dimensioning results together with the overall rollout plan of the customer, with regular updates in order to avoid unexpected expansions of the network due to higher-than-expected growth rates; or, on the opposite side, to avoid the deployment of excessive capacity for the true traffic demand.

B.5 MMS PERFORMANCE MONITORING AND OPTIMISATION

MMS application performance is measured in a number of ways: subscribers' complaints, network statistics and static/drive test measurements all provide a good starting point for analysis.

One of the tasks is to perform some measurements, using test tools capable of generating automated MMS traffic, to provide meaningful KPIs (key performance indicators) to accurately reflect the end-user experience. Then the current performance of the network can be evaluated and any further action decided on. Data and statistics from the NMS (network management system) can also be used as indicators of network performance and MMS service performance.

Another point is to identify current problem areas as well as potential bottlenecks in the delivery of MMS services as traffic increases. The network design (planning and dimensioning) can be reviewed with an emphasis on core and radio network configurations and dimensioning in line with the rollout plan. Consideration is also given to the interaction of MMS with other data applications as well as roaming and interoperability.

Then, a comprehensive bearer investigation can be conducted to determine the performance of the main MMS traffic bearers, for instance the GPRS network. The key point of the bearer investigation is to simulate the end users and generate real traffic, including tracing and post-processing at all layers of the radio and core protocol stacks and at all points in the end-to-end network. Static tests determine the core network performance for MMS as well as other data services. Dynamic measurements verify the impact of mobility management.

Depending upon the outcome of the bearer investigation, optimisation of the bearer itself may be needed. This consists of a detailed audit of both the packet core and radio networks, consisting of parameter consistency checks and detailed analysis of performance statistics.

Recommended parameter settings are generated with follow-up verification of performance improvements.

Once the MMS design is reviewed (including planning, dimensioning and configuration) and the bearer is optimised, MMS performance should be improved. Performance improvements can be checked by running some basic testing and comparing the results with the initial performance information.

B.6 CONCLUSION

MMS is the natural evolution of personal messaging. The first technology, SMS, already included in the GSM standard, although not supported in its early stages, is a well-known success story. MMS provides the power of the image. Its strength lies in the fact that it uses existing technology and infrastructure. The possible weaknesses may come from unforecast growth and its influence in voice and data services. Hence, proper planning of the system and its capability may help to reduce the impact of unexpected behaviour, although systematic update and close follow-up are required. MMS it is often considered to require strong system integration capabilities, although adequate system planning should be always a priority. Not only dimensioning, but rather an integrated MMS network plan covering the full network, is required to ensure that the network is working as expected. Testing and auditing the GPRS network in a systematic survey with the MMS system will provide a health check of the network capabilities as well as important findings affecting performance.

Appendix C
Location-based Services

Johanna Kahkonen
Nokia Networks

C.1 INTRODUCTION

Location-based services (LBS) are additional value provided in a wireless network, and are based on the known location of the mobile device. To enable these services, location positioning methods are required. This appendix explains two different categorisations for location-based services and gives examples. The main positioning methods used in the GSM and WCDMA networks are covered.

From the end-user's point of view, location-based services provide a way to associate needed information with a geographical position. For example, address information can be complemented with navigation instructions. For mobile network operators, location-based services are one possible way to provide tailored services. As the competition gets tighter, more personalised services are a possible way to succeed.

C.2 SERVICE DIVISION INTO CATEGORIES

Location-based services are divided here into three major categories: emergency support, operator services and commercial services. The division is explained briefly in Figure C.1.

Emergency support refers to an automatic facility that provides information about the mobile phone (MS) location in every case when an emergency call is received. In the United

Fundamentals of Cellular Network Planning & Optimisation A.R. Mishra.
© 2004 John Wiley & Sons, Ltd. ISBN: 0-470-86267-X

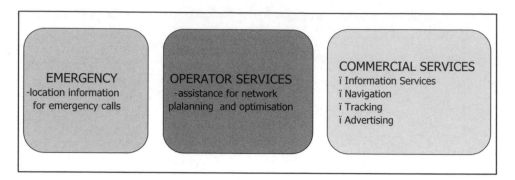

Figure C.1 Division of location-based services

States the positioning of an emergency call is regulated by the FCC (Federal Communications Commission). The accuracy of the given location estimate has quality requirements: it is set to 100 metres with 67% probability, and with 95% probability the accuracy needs to be better than 300 metres. This applies for the network-based LBS solution. For the mobile-phone-based solution the accuracy requirements are higher: 50 metres with 67% probability and 150 metres with 95% probability.

Operator services combine the location information and the network performance issues. To be able to increase network performance during the optimisation phase, it is important to investigate any problematic areas and locations in the network. However, the network management system (NMS) is able to provide network performance statistics with best accuracy of the cell level. This does not describe the actual position of interest, for example a location where a dropped call has happened. Location positioning methods enable higher accuracy to investigate geographically the problems in call events. Also, the geographical distribution of mobile phone users itself is interesting and useful information, especially when launching new services.

The commercial services can be targeted either to private users or to companies. Examples are information services, navigation, tracking and advertising.

Location-based information services provide for users desired information based on location. For example, yellow pages, news and traffic information can be delivered to the customer. The information delivery can be an instant answer to a query, such as "where is the nearest hotel?" Some information can be delivered repeatedly according to an agreement with the service provider and the customer, for example a weather forecast every morning.

With the location-based navigation services the type of information is similar, but the main issue is the navigation instructions. The information given is how to travel from the current location to a chosen point.

Location-based tracking can be directed to a person or a valuable object, to trace its movements. When considering a person to be traced, this service can be used in the same way as the safety telephone system, though the service is wireless. Tracking provides help for the relatives of a person who needs continuous support, such as an aged person or a child who needs medication. Tracking can be used, for example, by a car leasing company to trace its vehicles.

Location-based advertising delivers messages that are relevant and attractive according to the user's position. These messages are delivered to a person's mobile phone. The advertisement could, for example, show special offers in a department store that is close to the area where the mobile phone user currently is.

C.3 LOCATION POSITIONING METHODS

Positioning methods are based on three different types of technology: GPS-based, cell-coverage-based and triangular-based. CITARx and CI + RTT are coverage-based methods, while E-OTD and U-TDOA are examples of triangular-based methods. On the other hand, the technology behind positioning methods can be divided into three categories: network-based, mobile-phone-based and hybrid (a combination of network- and mobile-phone-based). In the following the basic methods for GSM and WCDMA networks are introduced.

The GPS (Global Positioning System) method relies on a GPS receiver integrated to a mobile phone, so this falls into the category of mobile-phone-based. The GPS method is accurate when finding the position of a mobile phone, but the downside is the mandatory GPS enabling feature in the MS. The device operability in GSM and WCDMA networks is determined by the mobile phone capabilities. The position of the mobile phone is estimated based on the measured distance between the GPS receiver and various satellites. The receiver needs to see a minimum of three satellites to be able to calculate a two-dimensional position. The more open is the view to the sky, i.e. the higher the number of satellites available, the better is the accuracy of the position estimate.

The A-GPS (Assisted-GPS) is a hybrid method, where an integrated GPS receiver in a mobile phone handles the position calculation and the network provides assistance data. LMU units are needed in the network side to handle the delivery of the assistance data. With A-GPS it is possible to increase the performance of the GPS receiver. Firstly, with the assistance data, the GPS receiver is able to locate the satellites much faster; and secondly, the positioning accuracy is higher. Also, fine-tuning of the location estimate is possible: after the first estimate, more assistance can be given.

The E-OTD (Enhanced-Observed Time Difference) method uses the signals received from at least three base stations to measure the difference in times they take from each BTS to the MS. The time difference is used as an input to triangular measurements to estimate the MS location. When the location is calculated in relation to the BTS locations, those coordinates have to be known by the system. Also the base stations have to be synchronised, and this is handled with LMU units, which compare the BTS time to absolute time. The MS has to have special software to be able to support the E-OTD method.

The U-TDOA (Uplink-Time of Arrival) method is in a sense the reverse of E-OTD. In U-TDOA the MS signal is measured by at least three base stations, and as in E-OTD the measurements are done by the MS. The arrival time of the data sent by the MS is registered in each BTS. The time is compared to the absolute time to get the difference in arrival times, and from this the MS location can be calculated using triangular measurement. U-TDOA does not require any updates or changes to the MS. On the other hand, all base stations need to be able to know the absolute time, which requires additional hardware, LMU units, for each BTS.

The Cell ID (CI) method in a GSM network uses the coordinates of the serving BTS as the location estimate for the mobile phone. As this method is purely network-based, it does not require any additional features to the mobile phone, so it is economic. The CI method can be enhanced with two features, Timing Advance (TA) and Received Signal Level (Rx), which increase the accuracy of the location estimates. The TA feature utilises TA measurements, and Rx utilises the neighbouring cell measurements, more particularly the signal levels measured for the neighbouring cells. Both TA and Rx measurements are an existing feature of GSM networks, so TA and Rx measurement results are available without any additional effort. The CI method in WCDMA networks operates in the same way, using the cell identity of the serving cell to estimate the location of the MS. The 3G CI method can be enhanced with the Round Trip Time (RTT) feature. The RTT is also an existing feature of WCDMA networks. The accuracy of a cell-based method depends on the size of the cell coverage area; the smaller the cell size, the higher is the accuracy.

An example of GSM network architecture for location positioning methods is presented in Figure C.2. The necessary new network elements are Gateway Mobile Location Center (GMLC), Serving Mobile Location Center (SMLC) and Location Measurement Unit (LMU). The location-based services' applications communicate with the GMLC, so the first task for the GMLC is to receive a request to locate a mobile phone. The GLMC initiates the positioning procedure in the network, by routing the request to the appropriate SMLC. The actual location estimate is calculated in the SMLC, which also handles the coordination of resources needed in the positioning procedure. The SMLC can either be integrated with a BSC or be an independent network element. The LMU unit, which works under control of the SMLC, is shown in Figure B.2 to be connected to the BTS, but alternatively it could be

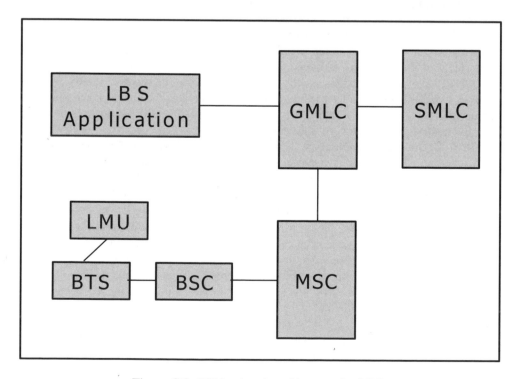

Figure C.2 GSM network architecture for LBS

connected to a BSC. Some of the location positioning methods require supportive measurements carried out by the LMUs. The GMLC finishes the positioning procedure by sending the estimated location to the external application.

C.4 ACCURACY

The accuracy of location positioning methods varies, as does the required accuracy for various location-based services. There is no existing system that is highly accurate yet at the same time inexpensive and technically not complex. Generally the GPS-based services are more accurate than network-based methods, as can be seen from Figure C.3.

With A-GPS it is possible to reach 3-metre accuracy, while with network-based methods the best possible accuracy is 100 metres. Note also that satellite- and network-based technologies differ in relation to their accuracy in different environments. The GPS systems require a clear view to the sky, so the best accuracy cannot be reached in dense city areas. Cell-based methods have the reverse accuracy behaviour, the highest accuracy being reached in city areas and the lowest in rural areas. This is because cell-based methods are dependent on the cell size.

Among the services introduced in the first section of this appendix, the highest accuracy is needed for navigation services. Very high accuracy is a requirement when giving detailed navigation instructions. Information services and advertising are not that sensitive with regard to positioning accuracy, so less complex systems can be used to provide them.

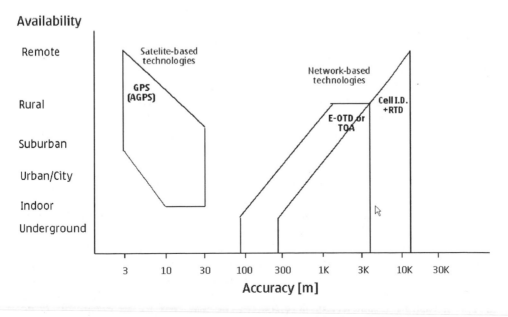

Figure C.3 Accuracy of the location positioning methods

C.5 NETWORK PLANNING ASPECTS

Network planning aspects are relevant for the network-based and hybrid location positioning methods. In the mobile-phone-based methods the location estimation is totally handled by the MS, so no changes to the network are needed. In addition to network planning, also some general SMLC and GMLC setting needs to be made to enable the location-based services in the network, and these settings affect also the mobile-phone-based methods. The settings are vendor-specific, so they are not described here.

The LBS network planning procedure consists of three steps:

- network planning for LBS

- implementation

- LBS operability and system accuracy verification.

In the first step the needed changes to the network are defined and planned. Then implementation means that the planned changes are transferred to the network. After implementation, the operability is tested and the positioning method accuracy is verified with measurements.

Only parameter planning is required for the CI-based positioning methods. The basic CI method is founded on serving-cell identity information and base station coordinates. The SMLC has to be able to find explicitly the correct cell, so the parameters enabling this need to be set to the SMLC. In GSM, in addition to CI, the cell identification parameters are for example LAC (Location Area Code) and BSIC (Base Station Identity Code). When considering the enhancing features for the CI methods, both GSM and WCDMA, there are more essential parameters. In this case the location estimate is no longer the coordinates of the base station and is common for all of the BTS sectors. More cell-level information needs to be given for the SMLC to increase the accuracy of the location estimation. The parameter settings in SMLC are vendor-specific (depending on the algorithms), but in the following can be found some ideas of relevant information both for CITARx and CI + RTT: antenna direction, antenna beam width, antenna height, BTS transmission power and terrain type.

For the A-GPS method, network planning includes on both LMU unit location planning and parameter planning for the assistance of data delivery and context. The LMU unit locations need to be planned so that the LMU coverage is adequate but the number of used LMU units is kept as low as possible. It is question of balancing the costs (hardware and installation costs) and the coverage (assistance availability). Parameter planning for A-GPS is necessary when defining the conditions for the assistance of delivery. If fine-tuning is used (i.e. the MS receives one location estimate to get more specific instructions to create a second and more accurate estimate), then also needed are cell parameters to decrease the area where the MS location can be searched.

The E-OTD method requires also LMU and parameter planning. Again the LMU planning has to be cost-efficient. The first planning rule for the E-OTD method is: the MS has to be able to receive signals from at least three different base stations to be able to calculate a location estimate. The second planning rule for E-OTD is: at least one LMU unit has to be serving in each location to be able to synchronise the base stations and calculate a location estimate. The basic parameters for E-OTD are the cell identification parameters, e.g. CI,

LAC and BSIC. The other parameters needed have to be verified from a system vendor, because these are dependent on algorithms.

The U-TDOA method does not require LMU planning, because this new network element is needed in every base station. The basic requirement for the SMLC parameters is similar to E-OTD, cell identification parameters.

Appendix D
End-to-End System Performance Measurement

N. B. Kamat
Nokia Networks

D.1 INTRODUCTION

With the maturing of the mobile phone industry networks in both developed and developing countries, the differentiator for operators is tending more towards network *quality*. Quality can mean a lot of different things to different network operators. The key common point among most operators is the 'user experience', which means end-to-end performance of the network. This is of increasing importance to operators since typical mobile networks now consist of a multitude of different components sourced from a vast range of suppliers. With IP or packet core services many vendors in the IT industry have become vendors to operators of mobile networks. The integration of all these different elements into one high-quality network has become a challenge for network planners worldwide.

Network planners are the equivalents of 'system designers' for wireless networks and they are often the ones measuring as well as optimising mobile networks. Network planners design the networks with given inputs of coverage, number of subscribers and call quality.

Since the mobile markets are maturing, the emphasis is now on the network quality and network quality testing. The end-user experience is sought to be captured in engineering terms as key performance indicators (KPIs). This appendix seeks to give a short overview of

Fundamentals of Cellular Network Planning & Optimisation A.R. Mishra.
© 2004 John Wiley & Sons, Ltd. ISBN: 0-470-86267-X

what these KPIs are, how they relate to user experience and how they are tested in practical networks.

D.2 BASIC SYSTEM KPIs

All the mobile networks have fundamental requirements of coverage and capacity.

Coverage

Coverage is normally expressed as location probability. This is usually given as a percentage. For example, a 95% outdoor location probability means that for 95% of the time, in 95% of the coverage area, the user will be able to make a call. Outdoor coverage can be measured by drive test equipment described in a later section. Such equipment uses some form of location equipment like global positioning system (GPS) and a test mobile phone attached to a laptop computer. Most of the testing is automated and the system is able to generate calls and then record a number of variables, coverage or signal strength in dBm being one of them.

The link budget normally gives the threshold value, which the coverage test must meet. The sensitivity of mobile phones being specified at -100 dBm for GSM 1800 band mobiles, this threshold is generally in the -95 dBm range. This represents the practical minimum signal strength below which the mobile cannot make a call.

Indoor coverage is far more difficult to measure, since this calls for a 'backpack' mounted test equipment and is far more labour-intensive since the speed is limited to that of a person walking in the buildings. Also, gaining access to buildings is normally a time-consuming affair. For these reasons, indoor coverage is approximated by an outdoor drive test. This is done by deciding an additional margin for 'building penetration loss', which could be anywhere from 10 to 25 dB depending on how much indoor coverage is required (one wall penetration, two walls, basement coverage, etc). Using a typical figure of 20 dB building penetration loss, the design criteria may be worded as '−95% location probability with a signal of −75 dBm'.

Capacity

The capacity of a network means how many subscribers the can network support. This is measured by 'network blocking'. Network blocking is expressed as a percentage. A 2% blocking means that for 2% of the time a user will not be able to get a call through. The network blocking can be measured both by drive tests as well as network statistics. In this case network statistics provide a more correct picture of the network blocking. For a second-generation GSM network, where the call set-ups are made on the SDCCH channel and traffic is carried on the TCH or traffic channels, it is possible for the user to get a signalling channel for call set-up but not a traffic channel. By counting the total number of SDCCH allocations versus traffic channel allocations, network statistics can give an accurate measurement of 'traffic channel blocking'. In many cases the government regulating authority, which allots the licences for wireless networks, specifies the blocking.

The signalling or SDCCH blocking can be similarly measured by statistics since the initial signalling is on the RACH (random-access channel). Network capacity can be increased by adding transmitters provided there is sufficient frequency available. The current 3G networks are also able to generate similar statistics to show traffic channel blocking as well as blocking on the call set-up.

Blocking percentages are normally used for voice calls but can be used for any circuit switched call like WCDMA video calls, which use a 64 kbps carrier.

Call Set-up Success Rate

Even though coverage and blocking measure the ability of the user to make a call, almost all operators measure Call Set-up Success Rate (CSSR) as a key performance indicator. The CSSR comes closest to the user experience in making a call. In addition to coverage and blocking, the CSSR takes into account the effect of interference, fading, etc., which affect the ability of the user to make a call. Typical mature networks with good coverage can have CSSR in the range 95–98%. Drive tests can be used to measure this KPI, but again network statistics give a more accurate picture.

For packet-switched calls the equivalent of call set-up success rate is the 'PDP context activation success rate' which is applicable for 2G, 2.5G and 3G networks.

It should be noted that the CSSR or PDP context activation success rates have to be adjusted for services purchased by the user. Typically many users do not purchase the packet call services but have phones capable of packet calls. Rejection of packet call requests from such users may form a high proportion of failures. In the circuit-switched call domain, pre-paid subscribers may be barred from certain services like roaming or international dialling by some operators. The effects of such failures need to be considered when calculating this KPI.

Dropped Call Rate

One characteristic of radio networks is a 'dropped call', which means the user experiences a termination of a call, which was not requested by either party in the call. Dropped calls can be due to coverage or due to interference and sometimes due to capacity. The number of dropped calls is measured by the system as network statistics and can be measured by drive testing to pinpoint the geographical location of the problem.

Different operators measure dropped call rates differently. Some consider dropped calls as those calls which dropped after the conversation started. The logic is that the user is not normally charged until the conversation is started. However, some operators charge users from the instant they initiate signalling. For such operators the appropriate dropped call definition is any call that drops after the user has attempted to make a call.

Mature 2G GSM networks can have dropped call rates of 1% for drops after the start of conversation and about 2% considering call set-ups.

Call Quality

Call quality measurements can be service-specific or independent of the service. 2G GSM systems provide a measure called 'Rx quality class' which varies from 0 to 7. Quality class 0 is the best and 7 is the worst. The quality classes are based on measurement of the bit error rates (BER) as measured by the error detection schemes built into the system. On a general level, voice quality of 0 with advanced codecs like EFR comes near to landline quality. Qualities of 2 and 3 have some perceptible distortions. Quality 4 has some amount of distortion. Conversation can still take place at Quality 5 levels depending on the ability of the people conversing. Quality classes 6 and 7 mean that no useful voice conversation is possible. In fact if there are too many instances of quality classes 6 and 7, system algorithms first try to hand over the call to a different cell or, in the worst case, the call may be dropped.

An example of the use of quality class would be to specify '99% of the network has quality class of between 0 to 5'. Again these parameters can be measured by drive testing which gives a good idea of the location of good and bad quality areas. However, some vendors are able to collect the statistics of quality as measured by mobiles and reported to the network. Use of such statistics gives a much more accurate picture of the call quality of the network.

A point to be noted is that the quality class measures the quality of the bearer and not of the service. A measure of service quality is provided in the sections on service quality.

Handover Success Rate

Modern wireless networks are called 'cellular' networks since they have small 'cells' each providing service in its immediate surrounding area. For the user to experience a seamless call, the user call is 'handed over' to the next cell when the user reaches the cell boundary. An extensive measurement and signalling protocol ensures that this handover is smooth and unnoticeable by users.

In spite of all precautions, handovers between cells fail. Handover failure can be a major cause of quality degradation in service or may lead to drop calls. Owing to its importance, the handover success rate is often specified as a key performance indicator. It must be noted, however, that a handover failure does not automatically mean a dropped call or bad quality. Most networks have a fallback mechanism where the call continues on the old cell and a new handover attempt is made after some waiting time. Handovers can also be triggered by other variables, such as capacity, call quality and mobile speed.

A typical handover success rate for mature 2G networks is of the order of 95% as measured from network statistics.

Call Set-up Time

The call set-up time is the time it takes for the user to get into a call. Again different operators worldwide define this differently. A common requirement is that call set-up time is the time measured from when the user presses the 'dial' button to when the ring tone is heard. Call set-up time can be measured by both drive tests or by the system itself. Owing to the current trend of having different vendors supplying different sub-systems (e.g. the radio and the switching parts supplied by different vendors), drive testing seems to be the test relied on for this.

To improve accuracy of measurement and to remove ambiguities, the call set-up time is defined as 'the interval between the dial command to the mobile and the receiving of the layer-three message '*alerting*' for 2G GSM networks. Similar definitions can be arrived at for other 2G and 3G systems.

Current trends seem to indicate a call set-up time of 4 seconds for a mobile to PSTN call and about 8 seconds for a mobile-to-mobile call. The mobile-to-mobile call takes longer since the called or 'B party' mobile has to be authenticated by the network.

The call set-up time depends heavily on the type and amount of authentication used on the network. Call set-up times are faster by at least 2–3 seconds where authentication is not used.

D.3 SERVICE-SPECIFIC KPIs

Voice Call Quality

It was noted in the early days of 2G networks that the quality of the bearer did not necessarily indicate the quality of the service. This was due to the error-correction scheme built into the coding schemes. Since voice was the first service to be offered in mobile networks, voice call quality has received a lot of attention.

In seeking a measure of quality that could measure the voice quality, *as perceived by the human ear*, a measure was developed called the 'mean opinion score' or MOS. In its 'pure' form this consists of a panel of humans who are asked to score the voice quality of a sample recorded from a mobile phone. Since these are subjective tests, the results are dependent on the types of people listening and even on the language used for the testing. This kind of testing is very long as well as expensive to do. The MOS was then automated and measurement tools are available for measurement, which try to mimic the performance of the human ear.

Typically MOS has a score range of 1 to 5. The result of 5 is the best and 1 is the worst. Most 2G systems with the EFR (Enhanced Full Rate) codec can give a MOS score of between 3 to 4, depending on network conditions.

MOS testing can be done only with drive test equipment.

Packet Call Quality

The packet call is characterised by two measurements: the latency and the throughput. Throughput can be described as the size of the data pipe available to the user, and latency can be described as the speed at which the data flow can start.

Throughput

Throughput is a measure of the size of data that can be transferred to and from the mobile user. Throughput is typically measured by doing an FTP (file transfer protocol) transfer of a file over the network and then working out the time taken. Typical file size for testing used is 1 Mb. This large file size is used so that the effects of cell changes are averaged out.

Throughput depends heavily on the capability of the mobile to use multiple time slots in 2G networks and using different sized bearers in 3G. Hence throughput is specified and measured per time slot in 2G and 2.5G networks and per bearer in 3G networks.

Typical 2G networks can support 10 kbps per time slot for the basic types of modulation. 2.5G networks are expected to support 80–100 kbps per time slot. WCDMA 3G networks have been specified with bearers of 384 kbps in the downlink, and performances of 330 kbps have been achieved in commercial networks.

Latency

One of the most frequent uses of a mobile packet network seems to be Internet surfing. This application typically sends out short packets to the network and then gets a bigger amount of data back from the network. Latency is the measure of how fast the system can transport the small amounts of data from the mobile to a specified server.

Latency in typically measured by 'pings'. A ping is a short message to a server by which the 'round trip time' can be measured. A typical round-trip time in 2G networks is 700 ms, and about 500 ms for 32-byte pings to servers located within the mobile network.

Video Call Quality

Currently there is no standard test available for testing the quality of a multimedia call, whether it be a still picture (MMS – multimedia messaging service) or a video call. Since all video codecs use some kind of compression, a standard is needed to evaluate the quality of a video call.

It should be possible to generate the equivalent of the MOS for a video call, which would have humans evaluating the quality of the call.

D.4 DRIVE TEST EQUIPMENT

Drive test equipment consists of a test mobile connected to a laptop computer, which has some kind of geographic, or location indication instrument attached to it. By default this is usually the global positioning system (GPS). The accuracy of commercial instruments based on this is usually of the order of 5 metres. The GPS system depends on the visibility of satellites to the receiving antenna. This can be difficult in heavily built-up areas (called 'urban canyons') and in places with a lot of trees. Also road tunnels, underpasses, bridges, etc., can cause such instruments to lose positioning information. For this purpose a 'dead reckoning' system can be fitted to the car, which basically connects to the pulses generated by the vehicle odometer.

Special testing software can be programmed to generate calls of a fixed duration and fixed repetition to a certain test number. The laptop then records all the measurements made by the phone, which are usually signal strength, neighbour lists, call quality and mobile output power. Most drive test equipments are also able to record the signalling messages between the mobile and the network.

Most of the KPIs mentioned above can be measured using such drive test equipment. In the initial stages of the network, especially before commercial launch, the drive test equipment is the only way to test the network since there are no other users on the network.

Drive test equipment is able to pinpoint the geographical location of any problem. The data collected can then be analysed to find out the cause and correct the problem. A second drive test can confirm whether the problem is corrected.

For indoor measurement the same set-up can be used. As GPS-based positioning does not work inside buildings, there is also available a complete test equipment in a single handset. This can be used to collect information about calls and the radio environment, but the user has to separately record position information based on markers than can be inserted while performing the test.

The advantage of drive testing is that precise geographical information is available and the testing does not depend on the vendor supplying the equipment. Further, some tests like MOS tests can be performed only with drive test equipment. Disadvantages are that drive testing is expensive and is limited to the roads, or in the case of indoor testing to accessible areas.

D.5 NETWORK STATISTICS

Most network vendors have counters to count the events occurring on the network. These counters are then collected by the network management system (NMS) for the particular sub-system. In a typical multivendor solution, the statistics from different vendors' equipment are collected into a special reporting tool to give an idea of the performance of the system as a whole.

Network element counters have not been standardised, so both the amount of counters available and their trigger points differ widely among vendors. For example, for 2G GSM radio systems one vendor has only about fifty counters available to the user while another has close to a thousand.

When using such counters from a different vendor, the user has to be aware of the differences in counter availability as well as their trigger points. In a typical multivendor scenario it is not unusual to have different values reported for the same KPI depending on the sub-system used for the reporting.

Network statistics gives a far more accurate and comprehensive picture of network performance since it includes all geographical area covered. Network statistics are collected automatically by the system and hence no special costs are associated with the measurement. The disadvantage of network statistics is that precise geographical information is missing. The problem can be localised down to cell level, and in some vendors' cases to distance from the cell antenna (based on timing advance). Also, user experience tests like MOS or the future video call-quality tests cannot be performed using network statistics.

The best solution for testing networks is to use a combination of both the tools. A typical application would be to use statistics to spot problems and localise them to cell level, and then use a drive test to get the details of the problem.

Appendix E
Erlang B Tables

Nezha Larhissi

Nokia Networks

An Erlang B or C table presents the estimated traffic to be handled by a certain number of TCHs according to a certain blocking probability. Since Erlang C admits queuing, more traffic can be handled for the same TCH/Probability combination than by Erlang B. However, in GSM standard, we use Erlang B tables with a common probability of usually 2% blocking. This table can be used, especially in dimensioning phases, whether to find the number of TRXs/Sites to be implemented for handling a certain traffic (based on some marketing inputs) or to calculate the amount of traffic expected in some areas depending on the number of TRXs/Sites. In the following table, N stands for the number of subscribers and B stands for (percent) blocking.

Fundamentals of Cellular Network Planning & Optimisation A.R. Mishra.
© 2004 John Wiley & Sons, Ltd. ISBN: 0-470-86267-X

N/B(%)	0.1	0.2	0.3	0.4	0.5	1	1.5	2	2.5	3	3.5	4	4.5	5
1	0	0	0	0	0.01	0.01	0.01	0.02	0.03	0.03	0.04	0.04	0.05	0.05
2	0.05	0.06	0.08	0.09	0.1	0.15	0.19	0.23	0.25	0.28	0.31	0.33	0.36	0.38
3	0.2	0.25	0.29	0.32	0.35	0.46	0.54	0.6	0.66	0.71	0.77	0.81	0.86	0.9
4	0.44	0.53	0.6	0.65	0.7	0.87	0.99	1.09	1.18	1.26	1.33	1.4	1.46	1.53
5	0.76	0.9	0.99	1.07	1.13	1.36	1.53	1.66	1.77	1.87	1.97	2.06	2.14	2.22
6	1.15	1.33	1.45	1.54	1.62	1.91	2.11	2.28	2.42	2.54	2.66	2.76	2.86	2.96
7	1.58	1.8	1.95	2.06	2.16	2.5	2.74	2.93	3.1	3.25	3.39	3.51	3.63	3.74
8	2.05	2.31	2.48	2.62	2.73	3.13	3.4	3.63	3.82	3.99	4.14	4.28	4.42	4.54
9	2.56	2.85	3.05	3.21	3.33	3.78	4.09	4.34	4.56	4.75	4.92	5.08	5.23	5.37
10	3.09	3.43	3.65	3.82	3.96	4.46	4.81	5.09	5.32	5.53	5.72	5.9	6.06	6.21
11	3.65	4.02	4.27	4.45	4.61	5.16	5.54	5.84	6.1	6.33	6.54	6.73	6.91	7.08
12	4.23	4.64	4.9	5.11	5.28	5.88	6.29	6.61	6.9	7.14	7.36	7.57	7.77	7.95
13	4.83	5.27	5.56	5.78	5.96	6.61	7.05	7.4	7.7	7.97	8.21	8.43	8.64	8.83
14	5.45	5.92	6.23	6.47	6.66	7.35	7.83	8.2	8.52	8.8	9.06	9.3	9.52	9.73
15	6.08	6.58	6.91	7.17	7.37	8.11	8.61	9.01	9.35	9.65	9.92	10.18	10.41	10.63
16	6.72	7.26	7.61	7.88	8.1	8.87	9.4	9.83	10.19	10.51	10.8	11.06	11.31	11.55
17	7.38	7.95	8.32	8.6	8.83	9.65	10.21	10.66	11.03	11.37	11.67	11.95	12.21	12.46
18	8.04	8.64	9.03	9.33	9.58	10.44	11.02	11.49	11.89	12.24	12.56	12.85	13.12	13.38
19	8.72	9.35	9.76	10.07	10.33	11.23	11.85	12.33	12.75	13.12	13.45	13.76	14.04	14.31
20	9.41	10.07	10.5	10.82	11.09	12.03	12.67	13.18	13.62	14	14.34	14.67	14.96	15.25
21	10.11	10.79	11.24	11.58	11.86	12.84	13.51	14.04	14.49	14.88	15.25	15.58	15.89	16.19
22	10.81	11.53	11.99	12.34	12.63	13.65	14.35	14.9	15.36	15.78	16.15	16.5	16.83	17.13
23	11.52	12.27	12.75	13.11	13.42	14.47	15.19	15.76	16.25	16.68	17.07	17.43	17.76	18.08
24	12.24	13.01	13.51	13.89	14.2	15.29	16.04	16.63	17.13	17.58	17.98	18.35	18.7	19.03
25	12.97	13.76	14.28	14.67	15	16.12	16.89	17.5	18.02	18.48	18.9	19.29	19.65	19.98
26	13.7	14.52	15.06	15.46	15.79	16.96	17.75	18.38	18.92	19.39	19.82	20.22	20.59	20.94
27	14.44	15.28	15.84	16.25	16.6	17.8	18.62	19.27	19.82	20.3	20.75	21.16	21.54	21.9
28	15.18	16.05	16.62	17.05	17.41	18.64	19.48	20.15	20.72	21.22	21.68	22.1	22.49	22.87
29	15.93	16.83	17.41	17.85	18.22	19.49	20.35	21.04	21.62	22.14	22.61	23.04	23.45	23.83
30	16.68	17.61	18.2	18.66	19.03	20.34	21.23	21.93	22.53	23.06	23.55	23.99	24.41	24.8
31	17.44	18.39	19	19.47	19.85	21.19	22.1	22.83	23.44	23.99	24.48	24.94	25.37	25.77
32	18.21	19.17	19.8	20.28	20.68	22.05	22.98	23.72	24.36	24.92	25.42	25.89	26.33	26.74
33	18.97	19.97	20.61	21.1	21.51	22.91	23.87	24.63	25.27	25.84	26.36	26.84	27.29	27.72
34	19.74	20.76	21.42	21.92	22.34	23.77	24.75	25.53	26.19	26.78	27.31	27.8	28.26	28.7

N/B(%)	0.1	0.2	0.3	0.4	0.5	1	1.5	2	2.5	3	3.5	4	4.5	5
35	20.52	21.56	22.23	22.75	23.17	24.64	25.64	26.44	27.11	27.71	28.26	28.76	29.23	29.68
36	21.29	22.36	23.05	23.57	24.01	25.51	26.53	27.35	28.04	28.55	29.2	29.72	30.2	30.66
37	22.08	23.16	23.87	24.41	24.85	26.38	27.42	28.25	28.96	29.59	30.16	30.68	31.17	31.64
38	22.86	23.98	24.69	25.24	25.69	27.25	28.32	29.17	29.89	30.52	31.1	31.64	32.15	32.62
39	23.65	24.79	25.52	26.08	26.53	28.13	29.21	30.08	30.81	31.47	32.06	32.61	33.12	33.61
40	24.45	25.6	26.35	26.92	27.38	29.01	30.11	31	31.75	32.41	33.02	33.58	34.1	34.6
41	25.24	26.41	27.18	27.76	28.23	29.89	31.02	31.91	32.68	33.36	33.97	34.54	35.08	35.58
42	26.04	27.24	28.01	28.6	29.08	30.77	31.92	32.84	33.61	34.31	34.93	35.51	36.06	36.57
43	26.84	28.06	28.84	29.45	29.94	31.66	32.83	33.76	34.55	35.25	35.89	36.48	37.04	37.57
44	27.64	28.88	29.68	30.29	30.8	32.54	33.73	34.68	35.49	36.2	36.85	37.46	38.02	38.56
45	28.45	29.71	30.52	31.15	31.66	33.43	34.64	35.61	36.43	37.15	37.82	38.43	39.01	39.55
46	29.25	30.54	31.37	32	32.52	34.32	35.55	36.53	37.37	38.11	38.78	39.4	39.99	40.54
47	30.07	31.37	32.21	32.85	33.38	35.21	36.47	37.46	38.31	39.06	39.75	40.38	40.97	41.54
48	30.88	32.2	33.06	33.71	34.25	36.11	37.38	38.39	39.25	40.02	40.71	41.36	41.96	42.54
49	31.69	33.04	33.91	34.57	35.11	37	38.29	39.32	40.2	40.97	41.68	42.34	42.95	43.53
50	32.51	33.88	34.76	35.43	35.98	37.9	39.21	40.25	41.15	41.93	42.65	43.32	43.94	44.53
51	33.33	34.72	35.61	36.29	36.85	38.8	40.13	41.19	42.09	42.89	43.62	44.3	44.93	45.53
52	34.15	35.56	36.47	37.16	37.73	39.7	41.05	42.12	43.04	43.85	44.59	45.28	45.92	46.53
53	34.98	36.4	37.32	38.02	38.6	40.6	41.97	43.06	43.99	44.81	45.56	46.26	46.91	47.53
54	35.8	37.25	38.18	38.89	39.47	41.5	42.89	44	44.94	45.78	46.54	47.24	47.91	48.54
55	36.63	38.09	39.04	39.76	40.35	42.41	43.82	44.93	45.89	46.74	47.51	48.23	48.9	49.54
56	37.46	38.94	39.9	40.63	41.23	43.31	44.74	45.87	46.85	47.7	48.49	49.21	49.89	50.54
57	38.29	39.79	40.76	41.5	42.11	44.22	45.67	46.82	47.8	48.67	49.46	50.2	50.89	51.55
58	39.12	40.65	41.63	42.38	42.99	45.13	46.59	47.76	48.75	49.63	50.44	51.18	51.89	52.55
59	39.96	41.5	42.49	43.25	43.87	46.04	47.52	48.7	49.71	50.6	51.42	52.17	52.88	53.56
60	40.79	42.35	43.36	44.13	44.76	46.95	48.45	49.65	50.66	51.57	52.4	53.16	53.88	54.57
61	41.63	43.21	44.23	45	45.64	47.86	49.38	50.59	51.62	52.54	53.37	54.15	54.88	55.57
62	42.47	44.07	45.1	45.89	46.53	48.77	50.31	51.53	52.58	53.51	54.35	55.14	55.88	56.58
63	43.31	44.93	45.97	46.76	47.42	49.69	51.24	52.48	53.54	54.48	55.33	56.13	56.88	57.59
64	44.15	45.79	46.84	47.65	48.31	50.6	52.17	53.43	54.5	55.45	56.32	57.12	57.88	58.6
65	45	46.65	47.72	48.53	49.2	51.52	53.11	54.38	55.46	56.42	57.3	58.11	58.88	59.61
66	45.84	47.51	48.59	49.41	50.09	52.43	54.04	55.33	56.42	57.39	58.28	59.11	59.88	60.62
67	46.69	48.38	49.47	50.3	50.98	53.35	54.98	56.28	57.38	58.37	59.27	60.1	60.88	61.63
68	47.54	49.24	50.35	51.18	51.87	54.27	55.92	57.22	58.34	59.34	60.25	61.09	61.89	62.64

(continued)

N/B(%)	0.1	0.2	0.3	0.4	0.5	1	1.5	2	2.5	3	3.5	4	4.5	5
69	48.39	50.11	51.22	52.07	52.77	55.19	56.85	58.18	59.31	60.31	61.24	62.09	62.89	63.65
70	49.24	50.98	52.1	52.96	53.66	56.11	57.79	59.13	60.27	61.29	62.22	63.08	63.9	64.67
71	50.09	51.85	52.98	53.85	54.56	57.04	58.73	60.08	61.24	62.27	63.21	64.08	64.9	65.68
72	50.94	52.72	53.87	54.74	55.46	57.96	59.67	61.03	62.2	63.24	64.19	65.08	65.9	66.69
73	51.8	53.59	54.75	55.63	56.35	58.88	60.61	61.99	63.17	64.22	65.18	66.07	66.91	67.71
74	52.65	54.46	55.63	56.52	57.25	59.8	61.55	62.95	64.14	65.2	66.17	67.07	67.91	68.72
75	53.51	55.34	56.52	57.42	58.15	60.73	62.49	63.9	65.11	66.18	67.16	68.07	68.92	69.74
76	54.37	56.21	57.4	58.31	59.05	61.66	63.44	64.86	66.08	67.16	68.14	69.06	69.93	70.75
77	55.23	57.09	58.29	59.2	59.96	62.58	64.38	65.81	67.04	68.13	69.14	70.06	70.94	71.77
78	56.09	57.96	59.18	60.1	60.86	63.51	65.32	66.77	68.01	69.11	70.12	71.06	71.95	72.78
79	56.95	58.84	60.07	61	61.76	64.44	66.27	67.73	68.98	70.1	71.11	72.06	72.96	73.8
80	57.81	59.72	60.95	61.9	62.67	65.36	67.21	68.69	69.95	71.08	72.11	73.06	73.96	74.82
81	58.67	60.6	61.85	62.79	63.57	66.29	68.16	69.65	70.92	72.06	73.1	74.06	74.97	75.84
82	59.54	61.48	62.74	63.69	64.48	67.22	69.1	70.61	71.89	73.04	74.09	75.06	75.98	76.86
83	60.4	62.36	63.63	64.59	65.39	68.15	70.05	71.57	72.87	74.02	75.08	76.06	76.99	77.87
84	61.27	63.24	64.52	65.5	66.29	69.08	71	72.53	73.84	75.01	76.08	77.07	78	78.89
85	62.14	64.13	65.41	66.4	67.2	70.02	71.95	73.49	74.81	75.99	77.07	78.07	79.01	79.91
86	63	65.01	66.31	67.3	68.11	70.95	72.89	74.45	75.79	76.98	78.06	79.07	80.03	80.93
87	63.87	65.9	67.21	68.2	69.02	71.88	73.85	75.41	76.76	77.96	79.05	80.07	81.04	81.95
88	64.74	66.78	68.1	69.11	69.93	72.81	74.79	76.38	77.73	78.94	80.05	81.08	82.05	82.97
89	65.61	67.67	69	70.01	70.84	73.75	75.74	77.34	78.71	79.93	81.04	82.08	83.06	83.99
90	66.48	68.56	69.9	70.92	71.76	74.68	76.7	78.31	79.69	80.92	82.04	83.09	84.07	85.01
91	67.36	69.44	70.8	71.82	72.67	75.62	77.65	79.27	80.66	81.9	83.03	84.09	85.09	86.04
92	68.23	70.33	71.69	72.73	73.58	76.56	78.6	80.23	81.64	82.89	84.03	85.09	86.1	87.06
93	69.1	71.22	72.59	73.64	74.49	77.49	79.55	81.2	82.61	83.87	85.03	86.1	87.11	88.08
94	69.98	72.11	73.49	74.55	75.41	78.43	80.5	82.17	83.59	84.86	86.02	87.11	88.13	89.1
95	70.85	73	74.39	75.46	76.33	79.37	81.46	83.13	84.57	85.85	87.02	88.11	89.14	90.12
96	71.73	73.89	75.3	76.36	77.24	80.31	82.41	84.1	85.55	86.84	88.02	89.12	90.16	91.15
97	72.61	74.79	76.2	77.27	78.16	81.24	83.37	85.07	86.53	87.82	89.02	90.12	91.17	92.17
98	73.48	75.68	77.1	78.19	79.07	82.18	84.32	86.04	87.5	88.81	90.01	91.13	92.19	93.19
99	74.36	76.57	78.01	79.09	79.99	83.12	85.28	87	88.48	89.8	91.01	92.14	93.2	94.22
100	75.24	77.47	78.91	80.01	80.91	84.07	86.24	87.97	89.46	90.79	92.01	93.15	94.22	95.24
101	76.12	78.37	79.81	80.92	81.83	85	87.19	88.94	90.44	91.78	93.01	94.16	95.23	96.26
102	77	79.26	80.72	81.83	82.75	85.95	88.15	89.91	91.42	92.78	94.01	95.16	96.25	97.29
103	77.88	80.16	81.63	82.75	83.67	86.89	89.11	90.88	92.41	93.76	95.01	96.17	97.27	98.31

N/B(%)	0.1	0.2	0.3	0.4	0.5	1	1.5	2	2.5	3	3.5	4	4.5	5
104	78.77	81.05	82.53	83.66	84.59	87.83	90.06	91.85	93.39	94.75	96.01	97.18	98.29	99.34
105	79.65	81.95	83.44	84.58	85.51	88.77	91.02	92.82	94.37	95.75	97.01	98.19	99.3	100.36
106	80.53	82.85	84.35	85.49	86.43	89.72	91.98	93.79	95.35	96.74	98.01	99.2	100.32	101.39
107	81.42	83.75	85.26	86.41	87.35	90.66	92.94	94.76	96.33	97.73	99.01	100.21	101.34	102.42
108	82.3	84.65	86.17	87.32	88.28	91.6	93.9	95.73	97.31	98.72	100.01	101.22	102.36	103.44
109	83.19	85.55	87.08	88.24	89.2	92.55	94.85	96.71	98.29	99.71	101.01	102.23	103.37	104.47
110	84.07	86.45	87.99	89.16	90.12	93.49	95.82	97.68	99.28	100.71	102.02	103.24	104.39	105.49
111	84.96	87.35	88.9	90.08	91.05	94.44	96.77	98.65	100.26	101.7	103.02	104.25	105.41	106.52
112	85.85	88.25	89.81	90.99	91.97	95.38	97.74	99.63	101.25	102.69	104.02	105.26	106.43	107.55
113	86.73	89.15	90.72	91.91	92.89	96.33	98.7	100.6	102.23	103.69	105.03	106.27	107.45	108.57
114	87.62	90.06	91.63	92.83	93.82	97.28	99.65	101.57	103.21	104.68	106.03	107.28	108.47	109.6
115	88.51	90.96	92.54	93.75	94.74	98.22	100.62	102.54	104.2	105.68	107.03	108.3	109.49	110.63
116	89.4	91.86	93.46	94.67	95.67	99.17	101.58	103.52	105.18	106.67	108.03	109.31	110.51	111.66
117	90.29	92.77	94.37	95.59	96.6	100.12	102.54	104.49	106.17	107.66	109.04	110.32	111.53	112.69
118	91.18	93.67	95.29	96.51	97.53	101.07	103.51	105.47	107.15	108.66	110.04	111.33	112.55	113.71
119	92.07	94.58	96.2	97.44	98.45	102.01	104.47	106.44	108.14	109.65	111.05	112.34	113.57	114.74
120	92.96	95.48	97.11	98.36	99.38	102.96	105.43	107.42	109.13	110.65	112.05	113.36	114.59	115.77
121	93.86	96.39	98.03	99.28	100.31	103.91	106.4	108.4	110.11	111.64	113.06	114.37	115.61	116.8
122	94.75	97.3	98.95	100.2	101.24	104.86	107.36	109.37	111.1	112.64	114.06	115.38	116.63	117.83
123	95.64	98.21	99.86	101.13	102.17	105.81	108.33	110.35	112.08	113.64	115.06	116.4	117.65	118.86
124	96.54	99.11	100.78	102.05	103.1	105.76	109.29	111.32	113.07	114.54	116.07	117.41	118.68	119.89
125	97.43	100.02	101.7	102.98	104.03	107.71	110.26	112.3	114.06	115.53	117.07	118.42	119.7	120.92
126	98.33	100.93	102.62	103.9	104.96	108.66	111.22	113.28	115.05	116.53	118.08	119.44	120.72	121.95
127	99.22	101.84	103.53	104.83	105.89	109.62	112.19	114.25	116.03	117.53	119.09	120.45	121.74	122.98
128	100.12	102.75	104.45	105.75	106.82	110.56	113.15	115.23	117.02	118.52	120.09	121.47	122.76	124.01
129	101.02	103.66	105.37	106.68	107.75	111.52	114.12	116.21	118.01	119.52	121.1	122.48	123.79	125.04
130	101.91	104.57	106.29	107.6	108.69	112.47	115.09	117.19	119	120.52	122.11	123.5	124.81	126.07
131	102.81	105.48	107.21	108.53	109.62	113.42	116.05	118.17	119.99	121.62	123.11	124.51	125.83	127.1
132	103.71	106.39	108.13	109.46	110.55	114.37	117.02	119.15	120.98	122.62	124.12	125.52	126.86	128.13
133	104.6	107.3	109.05	110.39	111.48	115.33	117.99	120.12	121.97	123.61	125.13	126.54	127.88	129.16
134	105.5	108.21	109.97	111.31	112.41	116.28	118.95	121.1	122.96	124.61	126.13	127.56	128.9	130.19
135	106.4	109.13	110.89	112.24	113.35	117.24	119.92	122.08	123.94	125.61	127.14	128.57	129.93	131.22
136	107.3	110.04	111.81	113.17	114.28	118.19	120.39	123.07	124.93	126.61	128.15	129.59	130.95	132.25
137	108.2	110.96	112.74	114.1	115.22	119.15	121.36	124.04	125.92	127.61	129.16	130.61	131.97	133.29
138	109.1	111.87	113.66	115.03	116.15	120.1	122.33	125.02	126.91	128.61	130.16	131.62	133	134.32

(continued)

N/B(%)	0.1	0.2	0.3	0.4	0.5	1	1.5	2	2.5	3	3.5	4	4.5	5
139	110	112.78	114.58	115.96	117.09	121.05	123.79	126	127.91	129.61	131.18	132.63	134.02	135.35
140	110.9	113.7	115.51	116.89	118.02	122.01	124.77	126.98	128.9	130.61	132.18	133.65	135.05	136.38
141	111.81	114.61	116.43	117.82	118.96	122.96	125.74	127.96	129.89	131.61	133.19	134.67	136.07	137.41
142	112.71	115.53	117.35	118.75	119.9	123.92	126.7	128.95	130.88	132.61	134.2	135.69	137.1	138.44
143	113.61	116.44	118.28	119.68	120.83	124.87	127.68	129.93	131.87	133.61	135.21	136.7	138.12	139.48
144	114.51	117.36	119.2	120.61	121.77	125.83	128.64	130.91	132.86	134.61	136.22	137.72	139.15	140.51
145	115.42	118.28	120.13	121.54	122.7	126.79	129.62	131.89	133.85	135.61	137.23	138.74	140.17	141.54
146	116.32	119.19	121.06	122.47	123.64	127.75	130.59	132.87	134.85	136.61	138.24	139.76	141.19	142.57
147	117.22	120.11	121.98	123.41	124.58	128.7	131.56	133.85	135.84	137.61	139.25	140.78	142.22	143.61
148	118.13	121.03	122.9	124.34	125.52	129.66	132.53	134.84	136.83	138.62	140.26	141.79	143.25	144.64
149	119.04	121.95	123.83	125.27	126.46	130.62	133.5	135.82	137.82	139.62	141.27	142.81	144.27	145.67
150	119.94	122.87	124.76	126.2	127.39	131.58	134.47	136.81	138.81	140.62	142.28	143.83	145.3	146.71
151	120.85	123.78	125.69	127.14	128.33	132.53	135.44	137.79	139.81	141.62	143.29	144.85	146.33	147.74
152	121.75	124.7	126.61	128.07	129.27	133.49	136.42	138.77	140.8	142.62	144.3	145.87	147.35	148.77
153	122.66	125.62	127.54	129.01	130.21	134.45	137.39	139.75	141.79	143.62	145.31	146.89	148.38	149.81
154	123.57	126.54	128.47	129.94	131.15	135.41	138.36	140.74	142.79	144.63	146.32	147.9	149.4	150.84
155	124.47	127.46	129.4	130.87	132.09	136.37	139.33	141.72	143.78	145.63	147.33	148.92	150.43	151.88
156	125.38	128.38	130.33	131.81	133.03	137.33	140.31	142.71	144.77	146.63	148.34	149.94	151.46	152.91
157	126.29	129.3	131.25	132.74	133.98	138.29	141.28	143.69	145.77	147.64	149.35	150.96	152.48	153.94
158	127.2	130.22	132.18	133.68	134.92	139.25	142.25	144.67	146.76	148.64	150.36	151.98	153.51	154.98
159	128.1	131.15	133.12	134.62	135.86	140.2	143.22	145.66	147.75	149.64	151.37	153	154.54	156.01
160	129.02	132.06	134.05	135.55	136.8	141.17	144.2	146.64	148.75	150.64	152.39	154.02	155.56	157.04
161	129.93	132.99	134.97	136.49	137.74	142.13	145.17	147.63	149.75	151.65	153.4	155.04	156.59	158.08
162	130.83	133.91	135.9	137.43	138.68	143.09	146.15	148.61	150.74	152.65	154.41	156.06	157.62	159.12
163	131.74	134.83	136.83	138.36	139.62	144.05	147.12	149.6	151.74	153.66	155.42	157.08	158.64	160.15
164	132.65	135.75	137.77	139.3	140.57	145.01	148.09	150.58	152.73	154.66	156.43	158.1	159.67	161.19
165	133.57	136.68	138.7	140.24	141.51	145.97	149.07	151.57	153.72	155.66	157.45	159.12	160.7	162.22
166	134.48	137.6	139.63	141.18	142.45	146.93	150.04	152.55	154.72	156.67	158.46	160.14	161.73	163.26
167	135.39	138.53	140.56	142.12	143.39	147.89	151.02	153.54	155.72	157.67	159.47	161.16	162.76	164.29
168	136.3	139.45	141.49	143.05	144.34	148.86	151.99	154.53	156.71	158.68	160.49	162.18	163.79	165.33
169	137.21	140.37	142.42	143.99	145.28	149.82	152.97	155.51	157.71	159.68	161.5	163.2	164.81	166.36
170	138.12	141.3	143.36	144.93	146.22	150.78	153.94	156.5	158.7	160.69	162.51	164.22	165.84	167.4
171	139.04	142.22	144.29	145.87	147.17	151.74	154.92	157.49	159.7	161.69	163.52	165.24	166.87	168.43
172	139.95	143.15	145.22	146.81	148.11	152.71	155.89	158.47	160.7	162.69	164.54	166.26	167.9	169.47
173	140.86	144.07	146.15	147.75	149.06	153.67	156.87	159.46	161.69	163.7	165.55	167.28	168.93	170.5

N/B(%)	0.1	0.2	0.3	0.4	0.5	1	1.5	2	2.5	3	3.5	4	4.5	5
174	141.77	145	147.09	148.69	150	154.63	157.85	160.44	162.69	164.7	166.56	168.3	169.96	171.54
175	142.69	145.92	148.02	149.63	150.95	155.6	158.82	161.43	163.69	165.7	167.58	169.33	170.99	172.58
176	143.6	146.85	148.96	150.57	151.89	156.56	159.8	162.42	164.68	166.72	168.59	170.35	172.01	173.61
177	144.52	147.78	149.89	151.51	152.84	157.52	160.78	163.41	165.68	167.72	169.6	171.37	173.04	174.65
178	145.43	148.7	150.83	152.45	153.78	158.49	161.75	164.39	166.63	168.73	170.62	172.39	174.07	175.68
179	146.35	149.63	151.76	153.39	154.73	159.45	162.73	165.38	167.67	169.73	171.63	173.41	175.1	176.72
180	147.26	150.56	152.7	154.33	155.68	160.42	163.71	166.37	168.67	170.74	172.65	174.43	176.13	177.76
181	148.18	151.49	153.63	155.27	156.62	161.38	164.69	167.36	169.67	171.75	173.66	175.46	177.16	178.79
182	149.09	152.41	154.57	156.21	157.57	162.35	165.66	168.35	170.67	172.75	174.68	176.48	178.19	179.83
183	150.01	153.34	155.5	157.16	158.52	163.31	166.64	169.34	171.66	173.76	175.69	177.5	179.22	180.87
184	150.92	154.27	156.44	158.1	159.47	164.28	167.62	170.32	172.66	174.77	176.7	178.52	180.25	181.9
185	151.84	155.2	157.38	159.04	160.41	165.24	168.6	171.31	173.66	175.77	177.72	179.54	181.28	182.94
186	152.76	156.13	158.31	159.98	161.36	166.21	169.57	172.3	174.66	176.78	178.73	180.57	182.31	183.98
187	153.68	157.06	159.25	160.92	162.31	167.17	170.55	173.29	175.66	177.78	179.75	181.59	183.34	185.02
188	154.59	157.99	160.19	161.87	163.25	168.14	171.53	174.28	176.66	178.79	180.76	182.61	184.37	186.05
189	155.51	158.91	161.12	162.81	164.2	169.1	172.51	175.27	177.65	179.8	181.78	183.63	185.4	187.09
190	156.43	159.84	162.06	163.76	165.15	170.07	173.49	176.26	178.65	180.81	182.79	184.66	186.43	188.13
191	157.35	160.78	163	164.7	166.1	171.03	174.47	177.25	179.65	181.82	183.81	185.68	187.46	189.17
192	158.27	161.71	163.94	165.64	167.05	172	175.45	178.24	180.65	182.82	184.83	186.7	188.49	190.2
193	159.18	162.64	164.87	166.59	168	172.97	176.43	179.23	181.65	183.83	185.84	187.73	189.52	191.24
194	160.1	163.56	165.81	167.53	168.95	173.93	177.41	180.22	182.65	184.34	186.86	188.75	190.55	192.28
195	161.02	164.5	166.75	168.48	169.89	174.9	178.39	181.21	183.55	185.35	187.87	189.77	191.58	193.32
196	161.94	165.43	167.69	169.42	170.85	175.87	179.37	182.2	184.55	186.35	188.89	190.8	192.61	194.35
197	162.86	166.36	168.63	170.37	171.79	176.83	180.35	183.19	185.55	187.36	189.9	191.82	193.64	195.39
198	163.78	167.29	169.57	171.31	172.74	177.8	181.32	184.18	186.65	188.37	190.92	192.84	194.68	196.43
199	164.7	168.22	170.51	172.26	173.7	178.77	182.31	185.17	187.65	189.88	191.94	193.87	195.71	197.47
200	165.62	169.15	171.45	173.2	174.64	179.74	183.29	186.16	188.65	190.89	192.95	194.89	196.74	198.51

Essential Reading

3GPP, 3G TR 23.923 ver. 3.0, 2000–05 *Combined GSM and Mobile IP Mobility Handling in UMTS IP CN*; 3G TS 22.796 ver. V2.0.0, 2000–06 *Study on Rel.2000 services and Capabilities.*

3GPP, Technical Specification Group 23.907, *Services and System Aspects, QoS Concepts.*

3GTS, 25.223 *Spreading and Modulation (TDD)*; 25.213, *Spreading and Modulation (FDD).*

Allen, K. C., "Observation of specific attenuation of millimetre waves by rain", *IEE A&P Conf.*, 1987.

Anfossi, D., Bacci, P. and Longhetti, A., "An application of Lidar technique to the study of nocturnal radiation inversion", *Atmos. Environ.*, **8**, 1960, 483–494.

Ansari and Evans, "Microwave propagation in sand and dust storms", IEE Proc. F, Communication, Radar and Signal Processing, 1982.

Bean et al, ESSA monograph, US Govt Printing Office, Washington, 1966.

Beck, R. and Panzer, H., "Strategies for handover and dynamic channel allocation in micro-cellular mobile radio systems", in *IEEE Vehicular Conf.*, 1989, 178–185.

Castro, J. P., *The UMTS Network and Radio Access Technology*, John Wiley & Sons, 2001.

Craig, K. H. and Kennedy, G. R., "Studies of microwave propagation on a microwave line-of-sight link", *IEE A&P Conf.*, 1987.

Crane, R. K., "Fundamental limitations caused by RF propagation", *Proc. IEEE*, **69**, 1981.

Fundamentals of Cellular Network Planning & Optimisation A.R. Mishra.
© 2004 John Wiley & Sons, Ltd. ISBN: 0-470-86267-X

Crane, R. K., "Prediction of attenuation due to rainfall on satellite systems", *Proc. IEEE*, **65**, 1977.

Doble, J., *Introduction to Radio Propagation for Fixed and Mobile Communication*, Artech House, 1996.

Effenberger, Strickland, Joy, "The effect of rain on a radome's performance", *Microwave Journal*, May 1986.

ETSI, GSM 04.60 *Digital Cellular Telecommunication System (Phase 2+): General Packet Radio Service (GPRS); Mobile Station (MS)–Base Station System (BSS) Interface; Radio Link Control/Medium Access Control Protocol*; GSM 05.05 *Digital Cellular Telecommunication System (Phase 2+): Radio Transmission and Reception*; GSM 03.30 *Digital Cellular Telecommunication System (Phase 2+): Radio Network Planning Aspects*.

Halonen, J. et al., *GSM, GPRS and EDGE Performance*, John Wiley & Sons, 2002.

Hata, M., "Emperical formula for propagation loss in land mobile radio services", *IEEE Trans. Vehicular Technology*, **VT-29**, 1980, 317–325.

Haykin, S., *Communication Systems*, John Wiley &Sons, 1983.

Heifiska Kari and Kangas, A., "Microcell propagation model for network planning", *Proceedings of PIMRC'96*, 1996, 148–152.

Holma, Harri, Toksala and Antti, *WCDMA for UMTS*, John Wiley & Sons, 2000.

Hunag, C. Y. and Yates, R. D., "Call admission control in cellular radio system", *IEEE Trans. Vehicular Technology*, **1**, 1992.

IETF, *TCP over 2.5G and 3G Wireless Networks*, IETF Internet-draft, October 2001.

ITU-R, "Okumara–Hata propagation model", in *Prediction Methods for the Terrestrial Land Mobile Services in the VHF and UHF bands*, ITU-R Recommendation P.529-2, ITU, Geneva, 1995, 5–7.

ITU-R, P.530-9 *Propagation Data and Prediction Methods Required for the Design of Terrestrial Line-of-Sight Systems*; P.837-3 *Characteristics of Propagation for Propagation Modelling*.

ITU-T, G.826 *Error Performance and Objectives for International, Constant Bit Rate Digital Paths at or above Primary Rate*; G.827 *Availability Parameters and Objectives for Path Elements of International Constant Bit Rate Digital Paths at or above the Primary Rate*.

Jakes, W. C. (ed.), *Microwave Mobile Communications*,Wiley-Interscience, 1974.

Janaswamy, R., *Radio Propagation and Smart Antennas for Wireless Communications*, Kluwer, 2000.

Kaaresoja, T. and Ruutu, J., "Synchronization and cell loss in cellular ATM evaluation system", *Proc. 5th Int. Workshop on Mobile Communication, moMuc'98*, Berlin, 12–14 October 1998.

Kazimierz Siwiak, *Radio Propagation for Antennas for Personal Communications*, Artech House, 1998.

Knorr, J. B., "Guided EM waves with atmospheric ducts", *Microwave and RF*, May 1985.

Ko, H., "A practical guide to anomalous propagation", *Microwave and RF*, April 1985.

Lee, W. C. Y., *Mobile Cellular Telecommunication Systems*, McGraw-Hill, 1990.

Lee, W. C. Y., *Mobile Communication Design Fundamentals*, John Wiley & Sons, 1993.

Lempiainen, J., *Radio Interface Planning for GSM/GPRS/UMTS*, Kluwer, 2001.

McAllister, L. G. et al., Acoustic sounding: a new approach to the study of atmospheric structure", *Proc. IEEE*, **57**, 1969, 579.

Mehrotra, A., *Cellular Radio Performance Engineering*, Artech House, 1994.

Mehrotra, A., *GSM System Engineering*, Artech House, 1997.

Mishra, Ajay R., "Cellular transmission network optimisation", *Advance* (Nokia Research Centre Journal), March 2001.

Mishra, Ajay R., "EDGE network analysis & optimisation", *WPMC*, Yokosuka, October 2003.

Mishra, Ajay R., "Transmission network planning and optimisation in third-generation access networks", *Advance* (Nokia Research Centre Journal), June 2003.

Mishra, Ajay R., "Observation of the fading phenomenon on the western coast of India, on a 7-GHz terrestrial path", *IMOC Conf.*, Rio de Janeiro, 1999.

Mitra, A. P. et al., "Tropospheric disturbance of 17–21 December 1974 and its effect on microwave propagation", *Boundary Layer Meteor*, **11**, 1977, 103.

Mouley, M. and Pautet, M.-B., *The GSM System for Mobile Communications*, 1992.

Oguchi, T., "Electromagnetic wave propagation and scattering in rain and other hydrometers", *Proc. IEEE*, **71**, 1983.

Ojanpera, T. and Prasad, R., *Wideband CDMA for Third Generation Mobile Communication*, Artech House, 1999.

Ojanpera, T., Prasad, R. and Harada, H., "Qualitative comparision of some multi-user detector algorithms for wideband CDMA", *Proc. IEEE Vehicular Tech. Conf.*, **1**, 1998, 46–50.

Okumara, Y., Ohmori, E., Kawano, T. and Fukuda, K., "Field strength and its variability in VHF and UHF land mobile radio service", *Review of Electrical Communication Laboratory*, **16**(9/10), 1968, 825–873.

Pajukoski, K. and Savusalo, J., "Wideband CDMA test system", *Proc. IEEE Int. Conf. on Personal Indoor and Mobile Radio Communication, PIMRC'97*, Helsinki, Finland, 1–4 September 1997, 669–672.

Parsons, D., *The Mobile Radio Propagation Channel*, Pentech Press, 1992.

Rummler, W. D., "A new selective fading model: application to propagation data", *BSTJ*, 1974.

Sarkar, S. K., *Radioclimatiological Effect on Tropospheric Radiowave Propagation over the Indian Sub-continent*, PhD thesis, University of Delhi, 1978.

Sipila, K., Laiho-Steffens, J., Wacker, A. and Jasperg, M., "Modelling the impact of the fast power control on the WCDMA uplink", *Proc. IEEE Vehicular Tech. Conf., VTC'99*, 1999, 1266–1270.

Soldani, D. and Abramowski, M., "An improved method for assessing the packet data transfer performance across an UMTS network", submitted WPMC 2002.

Tsurimi, H. and Suzuki, Y., "Broadband RF Stage architecture for software defined radio in handheld terminal application", *IEEE Communication Mag.*, **37**(2), 1999, 90–95.

UMTS, 22.01 *Service Aspects and Service Principles*; 22.25 *Quality of Service and Network Performance*; 22.05 *Service Capabilities*; 23.05 *Network Principles*.

Vigants, A., "Microwave obstructive fading", *BSTJ*, **60**, 1981.

Vigants, A., "Space diversity engineering", *BSTJ*, 1974.

Walfish, J. and Bertoni, H. L., "Theroretical model of UHF propagation in urban environments", *IEEE Trans. Antenna Propagation*, **AP-36**, 1988, 1788–1796.

Xia, H. H. et al., "Micro-cellular propagation characteristics for personal communication in urban and sub-urban environments", *IEEE Trans Vehicular Technology*, **43**, 1994, 743–752.

Yang, S. C., *CDMA RF System Engineering*, Artech House, 1998.

Zander, J., "Radio resource management: an overview", *Proc. IEEE Vehicular Technology Conf., VTC-96*, Atlanta, GA, 1996, 661–665.

Index

Fundamentals of Cellular Network Planning & Optimisation A.R. Mishra.
© 2004 John Wiley & Sons, Ltd. ISBN: 0-470-86267-X